Shaojian Fu
Mohammed Atiquzzaman

A Mobility Architecture for Terrestrial and Space Networks

Shaojian Fu
Mohammed Atiquzzaman

A Mobility Architecture for Terrestrial and Space Networks

Seamless IP-diversity-based Generalized Mobility Architecture (SIGMA)

VDM Verlag Dr. Müller

Impressum/Imprint (nur für Deutschland/ only for Germany)
Bibliografische Information der Deutschen Nationalbibliothek: Die Deutsche Nationalbibliothek
verzeichnet diese Publikation in der Deutschen Nationalbibliografie; detaillierte bibliografische
Daten sind im Internet über http://dnb.d-nb.de abrufbar.
Alle in diesem Buch genannten Marken und Produktnamen unterliegen warenzeichen-, marken-
oder patentrechtlichem Schutz bzw. sind Warenzeichen oder eingetragene Warenzeichen der
jeweiligen Inhaber. Die Wiedergabe von Marken, Produktnamen, Gebrauchsnamen,
Handelsnamen, Warenbezeichnungen u.s.w. in diesem Werk berechtigt auch ohne besondere
Kennzeichnung nicht zu der Annahme, dass solche Namen im Sinne der Warenzeichen- und
Markenschutzgesetzgebung als frei zu betrachten wären und daher von jedermann benutzt
werden dürften.

Coverbild: www.purestockx.com

Verlag: VDM Verlag Dr. Müller Aktiengesellschaft & Co. KG
Dudweiler Landstr. 125 a, 66123 Saarbrücken, Deutschland
Telefon +49 681 9100-698, Telefax +49 681 9100-988, Email: info@vdm-verlag.de
Zugl.: Norman, University of Oklahoma, Diss., 2005

Herstellung in Deutschland:
Schaltungsdienst Lange o.H.G., Zehrensdorfer Str. 11, D-12277 Berlin
Books on Demand GmbH, Gutenbergring 53, D-22848 Norderstedt
Reha GmbH, Dudweiler Landstr. 99, D- 66123 Saarbrücken
ISBN: 978-3-639-07958-6

Imprint (only for USA, GB)
Bibliographic information published by the Deutsche Nationalbibliothek: The Deutsche
Nationalbibliothek lists this publication in the Deutsche Nationalbibliografie; detailed
bibliographic data are available in the Internet at http://dnb.d-nb.de.
Any brand names and product names mentioned in this book are subject to trademark, brand or
patent protection and are trademarks or registered trademarks of their respective holders. The
use of brand names, product names, common names, trade names, product descriptions etc.
even without
a particular marking in this works is in no way to be construed to mean that such names may be
regarded as unrestricted in respect of trademark and brand protection legislation and could thus
be used by anyone.

Cover image: www.purestockx.com

Publisher:
VDM Verlag Dr. Müller Aktiengesellschaft & Co. KG
Dudweiler Landstr. 125 a, 66123 Saarbrücken, Germany
Phone +49 681 9100-698, Fax +49 681 9100-988, Email: info@vdm-verlag.de

Copyright © 2008 VDM Verlag Dr. Müller Aktiengesellschaft & Co. KG and licensors
All rights reserved. Saarbrücken 2008

Produced in USA and UK by:
Lightning Source Inc., 1246 Heil Quaker Blvd., La Vergne, TN 37086, USA
Lightning Source UK Ltd., Chapter House, Pitfield, Kiln Farm, Milton Keynes, MK11 3LW, GB
BookSurge, 7290 B. Investment Drive, North Charleston, SC 29418, USA
ISBN: 978-3-639-07958-6

Contents

List Of Tables

List of Illustrations

Chapter 1

Introduction

Wireless communication technology has undergone tremendous advance during recent years. Driven by the application and technology, mobile computing has become an increasingly important research area. New networking protocols, such as those dealing with IP mobility are indispensable for future computing paradigms.

Mobile IP (MIP) [1] is the standard proposed by Internet Engineering Task Force (IETF) to handle mobility of Internet hosts for mobile data communication. For example, it enables a TCP connection to remain alive and receive packets when a Mobile Host (MH) moves between points of attachment. MIP suffers from a number of drawbacks, and the most important issues of MIP identified to date are high handover latency and high packet loss rate.

Many improvements to Mobile IP have been recently proposed recently to reduce handover latency and packet loss, such as Mobile IPv6 (MIPv6) [2], Fast Handovers for Mobile IPv6 (FMIPv6) [3], Hierarchical Mobile IPv6 (HMIPv6) [4], and Fast and Hierarchical Mobile IPv6 (FHMIPv6) [4]. Even with these enhancements, Mobile IP can not completely solve the high latency problem, and the resulting packet loss rate is still high [5].

1.1 Motivation of SIGMA

As the amount of real-time traffic over wireless networks keeps growing, the deficiencies of the network layer-based Mobile IP, in terms of latency and packet loss, becomes more obvious. The question that naturally arises is: Can we find an alternative approach to network layer-based solutions for mobility support? Since most of the applications in the Internet are end-to-end, a transport layer mobility solution would be a natural candidate for an alternative approach. Recently, a number of transport layer mobility protocols have been proposed in the context of TCP, for instance, MSOCKS [6] and connection migration solution [7]. These protocols implement mobility as an end-to-end service without additional requirements on the network layer infrastructures; they are not aimed at reducing high latency and packet loss resulting from handovers. The handover latency for these schemes is in the scale of seconds.

Traditionally, various *diversity* techniques have been used extensively in wireless communications to combat channel fading by finding independent communication paths at the physical layer. Common diversity techniques include: space (or antenna) diversity, polarization diversity, frequency diversity, time diversity, and code diversity [8, 9]. Increasing numbers of mobile nodes are now equipped with multiple interfaces to take advantage of overlay networks (such as WLAN and GPRS) [10]. The development of Software Radio technology [11] also enables integration of multiple interfaces into a single network interface card. With the support of multiple IP addresses in one mobile host, a new form of diversity: *IP diversity* can be achieved.

A new transport protocol proposed by IETF, called Stream Control Transmission Protocol (SCTP), has recently received much attention from the research community [12]. In the field of mobile and wireless communications, the performance of SCTP over wireless links [13], satellite networks [14, 15], and mobile ad-hoc networks [16] is being studied. Multihoming is a built-in feature of SCTP,

2

which can be very useful in supporting IP diversity in mobile computing environments. Mobility protocols should be able to utilize these new hardware/software advances to improve handover performance.

The *objective* of this work is to design a new scheme for supporting low latency, low packet loss mobility called Seamless IP-diversity-based Generalized Mobility Architecture (SIGMA). The basic idea of SIGMA is to decouple location management from data transfer, and achieve seamless handover by exploiting IP diversity to keep the old path alive during the process of setting up the new path during handover. Although we illustrate SIGMA using SCTP, it is important to note that SIGMA can be used with other transport layer protocols that support IP diversity. It can also cooperate with normal IPv4 or IPv6 infrastructure without the support of Mobile IP.

1.2 Contributions of this Research

The contributions of the present research are as follows:

- Proposal and development of SIGMA, a network-friendly and seamless mobility architecture for IP hosts. Here "seamless" means low latency and low packet loss.

- Comparison of the performance of SIGMA with various MIPv6 enhancements including FMIPv6, HMIPv6 and FHMIPv6 using *ns*-2 simulation.

- Development of analytical models of handover performance, signaling cost, and survivability of SIGMA.

- Development of a hierarchical location management scheme for SIGMA, which is also applicable to other transport layer mobility solutions.

- Demonstration of the applicability of SIGMA to the management of satellite handovers in space networks.

1.3   Book Structure

The rest of this book is structured as follows: Chapter 2 introduces the motivation for IP mobility, and reviews recent literature in this area. Chapter 3 outlines the architecture of SIGMA. The handover performance of SIGMA and MIPv6 enhancements is compared by simulation and analytical model in Chapter 4. Signaling cost and survivability evaluation of SIGMA will be presented in Chapters 5 and 6, respectively. The hierarchical location management scheme for SIGMA is described in Chapter 7. Application of SIGMA to satellite handovers in space networks is discussed in Chapter 8. Interoperability between SIGMA and existing Internet security mechanisms is discussed in Chapter 9. Finally, concluding remarks are presented in Chapter 10.

Chapter 2

Overview of IP Mobility & Literature Review

In this chapter, we first introduce the concept of IP mobility in Sec. 2.1. The basic idea of Mobile IP, which implements IP mobility at the network layer, is described in Sec. 2.2. Then the recent research efforts in MIP and transport layer mobility are reviewed in Sec. 2.3 and 2.4, respectively.

2.1   Why Introduce IP Mobility?

In the current Internet, IP addresses are primarily used to identify particular end systems. In this respect, IP addresses are often thought of as being semantically equivalent to a Domain Name Server's Fully Qualified Domain Name (FQDN). In other words, one can (conceptually) use either an IP address or FQDN to identify one particular computer in the Internet. Popular transport protocols, such as Transmission Control Protocol (TCP) [17, 18], keep track of their internal session state between the communicating endpoints by using the IP address of the two endpoints and their port numbers. On the other hand, IP addresses are also used to find a route between the endpoints. The route does not have to be the same in both directions. Therefore, a mobile host needs to have a stable IP address in order to be uniquely identifiable to other Internet hosts. However, when a mobile host moves from one network to another, as shown in Fig. 2.1, the IP address of an MH will change due to the enforced hierarchical address structure of the Internet.

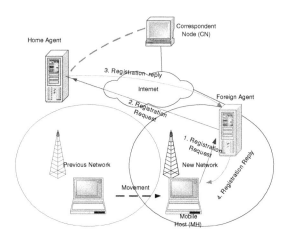

Figure 2.1: Mobile IP handover caused by MH movement.

If the MH has only standard IP stack in its operating system, the TCP connection will break up following this movement.

2.2 Basic Concepts of Mobile IP

Mobile IP [1, 19] is the standard proposed by IETF to offer seamless mobile computing. Mobile IP extends IP by allowing a mobile computer to utilize two IP addresses: one for identification, and another for routing. For example, it enables a TCP connection to keep alive and re-route packets when the mobile host moves between points of attachment.

Mobile IP defines a number of terms [1] to support IP host mobility; here we list the five basic ones:

- Mobile Host (MH): A host or router that changes its point of attachment from one network or subnetwork to another, without changing its IP address.

A mobile node can continue to communicate with other Internet nodes at any location using its (constant) IP address.

- Correspondent Node (CN): The host that has a transport layer connection with MH to perform data communication. CN can be any host on the Internet.

- Home Agent (HA): A router on a mobile node's home network that delivers packets to moved mobile nodes, and maintains current location information for each.

- Foreign Agent (FA): A router on a mobile node's visited network that cooperates with the home agent to complete the delivery of packets to the mobile node while it is away from home.

- Home Address: A long-term IP address for MH on its home network.

- Care of Address (CoA): Address for MH that reflects its current point of attachment when it is away from its home network.

In order to route data packets after MH has moved into its new location, MIP also defines simple mechanisms to deliver packets to the mobile node when it is away from its home network. Following each change of point of attachment, MH registers with HA with its new CoA. When HA receives IP packets for MH, it will encapsulate the packets with MH's CoA as the destination address and forward them to FA. FA will decapsulate the packets by strip off the outer IP header and deliver the packets to MH.

2.3 Recent Research on Improving Mobile IP

A number of improvements to Mobile IP have been proposed to reduce handover latency and packet loss. IP micromobility protocols, like Hierarchical IP [20], HAWAII [21] and Cellular IP [22], use hierarchical foreign agents to reduce the frequency and latency of location updates by handling most of the handovers locally. Low latency Handoffs in Mobile IPv4 [23] uses pre-registrations and post-registrations which are based on utilizing link-layer event triggers to reduce handover latency.

Optimized smooth handoff [24] not only uses a hierarchical FA structure, but also queues packets at the visited FA buffer, and forwards packets to MH's new location. To facilitate packet rerouting after handover and reduce packet losses, Jung et al. [25] introduce a location database that maintains the time delay between the MH and the crossover node. Mobile Routing Table (MRT) has been introduced at the home and foreign agents [26], and a packet forwarding scheme similar to optimized smooth handoff [24] is also used between FAs to reduce packet losses during handover. A Reliable Mobile Multicast Protocol (RMMP) [27] uses multicast to route data packets to adjacent subnets to ensure low packet loss rate during MH roaming. Fu et al. [28] use SCTP to improve the performance of MIP by utilizing SCTP's unlimited SACK GapAck Blocks.

MIPv6 [2] removes the concept of FA to reduce the requirement on infrastructure support (only HA required). Route Optimization is built in as an integral part of MIPv6 to reduce triangular routing problem encountered in MIPv4 [2]. FMIPv6 [3] aims to reduce handover latency by configuring a new IP address before entering a new subnet. This results in a reduction in the time required to prepare for new data transmission; the packet loss rate is thus expected to decrease. Like Hierarchical IP in MIPv4, HMIPv6 [4] also introduces a hierarchy of mobile agents to reduce the registration latency and the possibility of an outdated Collocated Care of Address (CCoA). FMIPv6 and HMIPv6 can be used together, as suggested by FHMIPv6 [4], to further improve the performance. The combination of Fast Handover and HMIPv6 allows performance improvements by taking advantage of both hierarchial structure and link layer triggers. However, like FMIPv6, FHMIPv6 also relies heavily on accurate link layer information. MH's high movement speed or irregular movement pattern may reduce the performance gains of these protocols. Even with the above enhancements, Mobile IP still can not completely remove the handover latency, resulting in a high packet loss rate [5].

2.4 IP Mobility at the Transport Layer

In Sec.2.3, we reviewed recent research efforts on Mobile IP, which implements IP mobility at the network layer. However, IP mobility can also be implemented at the transport layer. A number of transport layer mobility protocols have been proposed in the context of TCP, for example, MSOCKS [6] and connection migration solution [7].

2.4.1 MSOCKS

MSOCKS [6] is built around a proxy that is inserted in the communication path between a mobile node and its correspondent hosts. The architecture of MSOCKS is shown in Fig. 2.2. For each data stream from a mobile node to a correspondent host, the proxy is able to maintain one stable data stream to the correspondent host, isolating the correspondent host from any mobility issues. Meanwhile, the proxy can simultaneously make and break connections to the mobile node as needed to migrate data streams between network interfaces or subnets. Basically, the proxy in MSOCKS is conceptually equivalent to HA in Mobile IP except the proxy works at the transport layer while HA works at the network layer.

The proxy machine in MSOCKS breaks the transport layer connection into two parts, one from CN to proxy, and another from proxy to MH. The proxy functions have to be carefully tuned to preserve the end-to-end semantics. In addition to introducing modifications to Internet infrastructure, MSOCKS could suffer from scalability issues when the proxy manages a large number of mobile hosts, since all the data traffic sent to MHs needs to be processed at the proxy machine.

9

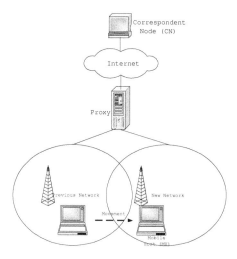

Figure 2.2: Architecture of MSOCKS.

2.4.2   TCP Connection Migration

A TCP connection is uniquely identified by a 4-tuple: source address, source port, destination address and destination port [18]. Packets addressed to a different address, even if successfully delivered to the TCP stack on the mobile host, must not be de-multiplexed to a connection established from a different address. Similarly, packets from a new address are also not associated with connections established from a previous address. This is crucial to the proper operation of servers on well-known ports. In TCP connection migration [7], a new proposed TCP option is called "*Migrate*", is included in SYN segments, that identifies a SYN packet as part of a previously established connection, rather than a request for a new connection. The timeline of TCP connection migration is shown in Fig. 2.3. The *Migrate* option contains a token that identifies a previously established connection on the same destination address/ port pair. The token is negotiated during initial

10

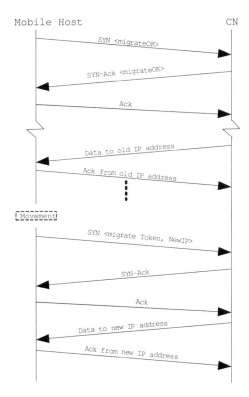

Figure 2.3: Timeline of TCP connection migration.

connection establishment through the use of a *Migrate-Permitted* option. After a successful token negotiation, TCP connections may be uniquely identified by either their traditional source address, source port, destination address, destination port 4-tuple, or a new source address, source port, token triple on each host.

A mobile host may restart a previously-established TCP connection from a new address by sending a special *Migrate* SYN packet that contains the token identifying the previous connection. The CN will then re-synchronize the connection with

MH at the new end point. A migrated connection maintains the same control block and state (with a different end point, of course), including the sequence number space; any necessary retransmissions can thus be requested in the standard fashion. This ensures preservation of true end-to-end semantics of TCP, even when MH has changed its point of attachment on the Internet.

2.5 Summary

In this chapter, the basic idea of Mobile IP, which implements IP mobility at the network layer, is described. MIP suffers from a number of drawbacks, and the most important issues of MIP are high handover latency and high packet loss rate. A number of improvements to Mobile IP have been proposed to improve its handover performance. However, even with the recent proposed enhancements, Mobile IP still can not completely remove the handover latency, resulting in a high packet loss rate.

IP mobility can also be implemented at the transport layer. MSOCKS and TCP connection migration solutions implement mobility as an end-to-end service without the requirement to change network layer infrastructures; however, they did not aim to reduce the high latency and packet loss resulting from handovers. As a result, the reported handover latency by these schemes is still in the scale of seconds, which is not acceptable for most real-time applications. Furthermore, important issues when developing mobility protocols in the Internet, such as signaling cost, scalability, survivability measures, and security issues have not been analyzed by previous work.

Chapter 3

Proposed Architecture: SIGMA

We have presented the concept of IP mobility, and also reviewed recent research efforts in the network layer and transport layer to implement IP mobility in Chapter 2. The previously proposed schemes can not completely remove the high handover latency and packet loss rate. In this chapter, we describe the architecture of SIGMA for IP mobility to achieve seamless IP handovers. The main idea of SIGMA is to decouple location management from data transfer, and achieve seamless handover by exploiting IP diversity to keep the old path alive during the process of setting up the new path during handover.

We illustrate SIGMA using SCTP; we therefore introduce the main features of SCTP in Sec. 3.1. SIGMA signaling procedure can be divided into five parts as are described in Sec. 3.3. The timing diagram and location management scheme used in SIGMA are described in Secs. 3.4 and 3.5, respectively.

3.1 Introduction to SCTP — A New Internet Transport Layer Protocol

Recent increase in interest in transmitting Voice over IP (VoIP) networks has led to the development by the IETF of a new transport layer protocol, called Stream Control Transmission Protocol (SCTP) [29], for the IP protocol suite. Although, the initial objective of developing SCTP was to provide a robust protocol for the transport of VoIP signalling messages over an IP network, later developments have also made it useful as a transport protocol for a wider range of applications, resulting in moving the standardization work of SCTP from SIGTRAN to the Transport Area Working Group (TSVWG) of IETF in February 2001.

SCTP is a reliable network-friendly transport protocol which can co-exist with TCP in the same network. The design of SCTP absorbed many of the strengths of TCP, such as window-based congestion control, error detection and retransmission, that led to its success during the explosive growth of the Internet. Moreover, SCTP incorporated several new features that are not available in TCP, which has made SCTP one of the hot topics in networking technology [12,30,31]. The main features of SCTP, multi-homing, multi-streaming, and congestion control, and difference between SCTP and TCP are described below.

3.1.1 SCTP Multi-homing

SCTP's multi-homing allows an association between two end points to span multiple IP addresses or network interface cards. An example of SCTP multi-homing is shown in Fig. 3.1, where both endpoints A and B have two interfaces bound to an SCTP association. The two end points are connected through two types of links: satellite at the top and ATM at the bottom. One of the links is designated as the primary while the other can be used as a backup link in the case of failure of the primary, or when the upper layer application explicitly requests the use of the backup. Retransmission of lost packets can also be done over the secondary address.

The built-in support for multi-homed endpoints by SCTP can utilize the network redundancy, and is especially useful in environments that require high-availability of the applications, such as SS7 signaling transport. A multi-homed SCTP association can speed up recovery from link failure situations without interrupting any ongoing data transfer.

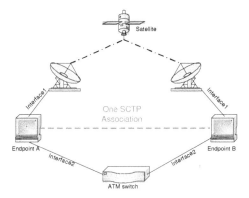

Figure 3.1: An SCTP association with multi-homed endpoints.

3.1.2 Multi-streaming

Multi-streaming allows data from the upper layer application to be multiplexed onto one channel (called an association in SCTP) as shown in Fig. 3.2. Sequencing of data is done within a stream; if a segment belonging to a certain stream is lost, segments (from that stream) following the lost one will be stored in the receiver's stream buffer until the lost segment is retransmitted from the source. However, data from other streams can still be passed to the upper layer application. This avoids the Head-Of-Line (HOL) blocking found in TCP, where a single stream carries data from all the upper layer applications. In other words, the HOL effect is limited within the scope of individual streams, but does not affect the entire association.

An example application of using SCTP multi-streaming in Web browsing is shown in Fig. 3.3. Here, an HTML page is split into five objects: a java applet, an ActiveX control, two images, and plain text. Instead of creating a separate connection for each object as in TCP, SCTP is making use of its multi-streaming feature to speedup the transfer of HTML pages. By transmitting each object in

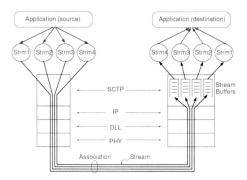

Figure 3.2: An SCTP association consisting of four streams.

a separate steam, the HOL effect between different objects can be eliminated. If one object is lost during the transfer, the others can still be delivered to the Web browser at the upper layer, while the lost object is being retransmitted from the Web server. This results in a better response time to users while using only one SCTP association for an HTML page.

3.1.3 SCTP Congestion Control

SCTP congestion control is based on the well proven rate-adaptive window-based congestion control scheme of TCP. This ensures that SCTP will reduce its sending rate during network congestion and prevent congestion collapse in a shared network. SCTP provides reliable transmission and detects lost, reordered, duplicated or corrupt packets. It provides reliability by retransmitting lost or corrupt packets. However, there are several major differences between TCP and SCTP as summarized below:

- SCTP incorporates a fast retransmit algorithm based on SACK gap reports similar to that of TCP. This mechanism speeds up loss detection and increases the bandwidth utilization. One of the major differences between

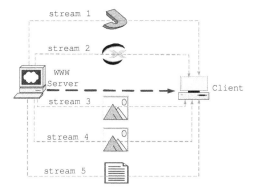

Figure 3.3: Multi-streaming in Web browsing.

SCTP and TCP is that SCTP doesn't have an explicit fast-recovery phase. SCTP achieves fast recovery automatically with the use of SACK [29].

- Compared to TCP, The use of SACK is mandatory in SCTP, which allows more robust reaction in the case of multiple losses from a single window of data. This avoids a time-consuming slow start stage after multiple segment losses, thus saving bandwidth and increasing throughput.

- During slow start or congestion avoidance of SCTP, the congestion window (*cwnd*) is increased by the number of acknowledged bytes; in TCP it is increased by the number of ACK segments received. Since the TCP sender increases the size of *cwnd* based on the number of arriving ACKs, the widely used delayed ACK will reducing the number of ACKs which in turn slows the *cwnd* growth rate.

- During congestion avoidance of SCTP, *cwnd* can only be increased when the full *cwnd* is utilized; this restriction does not exist in TCP.

17

- TCP begins fast retransmission after the receipt of three DupACKs; SCTP begins after four DupACKs. SCTP is able to clock out new data on receipt of the first three DupACKs and retransmit a lost segment by ignoring whether the flight size is less than *cwnd*; TCP can only begin data retransmission on the receipt of the third DupACK.

3.1.4 Difference between TCP and SCTP

The differences between TCP and SCTP are summarized in Table 3.1. The first three rows compare the messages exchanged during TCP connection/SCTP association setup & shutdown. The fourth and fifth rows of the table relate to the delivery of segments to the application at the receiver. The sixth row considers message boundaries after transmission by the transport layer protocols. The last two rows of Table. 3.1 relate to keep-alive messages. The differences revealed in the above comparison of the two transport layer protocols reflect understanding of the deficiencies of TCP by the research community during the past twenty years of practice.

3.1.5 Recent research work on SCTP

A large amount of work has been carried out in the last few years in evaluating the performance of SCTP [12]. For example, the co-existence study of SCTP and TCP in the Internet has shown that SCTP traffic has the same impact on the congestion control decision of TCP connections as normal TCP traffic [32]. The study on the effects of SCTP multi-homing in the recovery of SS7 network linkset failures has shown that the multi-homing feature of SCTP can help the endpoints to detect link failures earlier than the traditional approaches, and is also transparent to upper-layer applications [33]. Research on SCTP multi-streaming in reducing the latency of streaming multimedia in high-loss environments shows that multi-streaming results in a slower degradation in network throughput as the the loss

18

Table 3.1: Comparison of TCP and SCTP

Protocol	TCP	SCTP
Setup messages	three-way handshake	four-way handshake
Shutdown messages	four-way handshake	three-way handshake
Half-open support	supported	not supported
Ordered delivery	strict ordered	ordered within a stream
Unordered delivery	not supported	supported
Message boundary	no boundary stream-oriented	boundary preserved message-oriented
Multi-homing	not supported	supported
Multi-streaming	not supported	supported
SACK support	optional	mandatory
Keep-alive heart-beat	optional	mandatory
Heartbeat interval	≥ 2 hours	30 secs by default

rate increases [15, 34]. Moreover, user satisfaction is increased with the improved multimedia quality provided by this feature [34].

In the wireless networking area, the performance of SCTP in mobile networks [28] and wireless multi-hop networks [16] has been studied. The performance of SCTP in MIP was investigated by Fu et al. [28]; it was shown that the support of a large number of SCTP GapACK blocks [29] in its SACK chunks can expedite error discovery and lost packet retransmission, and result in better performance than TCP-Reno and TCP-SACK. Ye et al. [16] have shown that the throughput of an SCTP association degrades when the number of hops between the sender and receiver increases, mainly due to the hidden node and exposed node problems.

3.2   Architecture of SIGMA

The architecture of SIGMA is shown in Fig. 3.4. Both access routers in the previous domain and new domain are standard IP routers. The DHCP server can be used to provide IP address upon the request from MH when MH moves into the new domain. The access router in the new domain can also be combined together with the DHCP server. If IPv6 Stateless Address Auto-configuration (SAA) [35] is used for configuring new IP address for MH, no DHCP server is required.

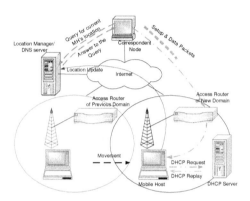

Figure 3.4: Architecture of SIGMA.

When MH stays in the overlapping region between the previous domain and the new domain, the MH maintains two IP address, one from the old domain and another from the new domain. MH achieves IP diversity by allowing CN to be able to send data packets to anyone of the two IP addresses. The handover latency and packet loss rate can be reduced by utilizing this IP diversity. The detailed handover procedure and handover timing diagram in SIGMA will be presented in Sec. 3.3 and 3.4, respectively.

A location manager need to be setup to record the current location of MH. This is required for new association setup request from new CNs (other than the ones MH is communicating with) to be delivered to MH. Following every change of point of attachment, MH need to update the location manager. CN first need to query the location manager for the current location of MH, then send association setup request or other data packets to MH. The standard DNS servers can be used as the location manager which maps MH's domain name to its current IP address. The detailed discussion of location management of SIGMA is presented in Sec. 3.5.

3.3 SIGMA Handover Process

A typical mobile handover in SIGMA, using SCTP as an illustration, is shown in Fig. 3.5, where MH is a multi-homed node connected to two wireless access networks. CN is a node sending traffic to MH, representing services like file download or web browsing by mobile users.

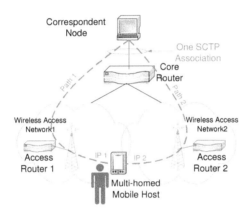

Figure 3.5: An SCTP association with multi-homed mobile host.

The handover process of SIGMA can be described by the following five steps.

STEP 1: Layer 2 handover and obtain new IP address

Referring to Fig. 3.5 as an example, the handover preparation procedure begins when MH moves into the overlapping radio coverage area of two adjacent subnets. In state of the art mobile system technologies, when a mobile host changes its point of attachment to the network, it needs to perform a Layer 2 (data link layer) handover to cutoff the association with the old access point and associate with a new one. As an example, in IEEE 802.11 Wireless Local Area Network

21

(WLAN) infrastructure mode, this Layer 2 handover will require several steps: detection, probe, and authentication and association with new AP. Only after these procedures have been finished, higher layer protocols can proceed with their signaling procedure, such as Layer 3 router advertisements. Once the MH finishes Layer 2 handover and receives the router advertisement from the new access router (AR2), it should begin to obtain a new IP address (IP2 in Fig. 3.5). This can be accomplished through several methods: Dynamic Host Configuration Protocol (DHCP), DHCPv6, or IPv6 SAA [35].

STEP 2: Add IP addresses into the association

Initially, when the SCTP association is setup, only the CN's IP address and MH's first IP address (IP1) are exchanged between CN and MH. After the MH obtains IP address IP2 in STEP 1, MH should bind IP2 also into the association (in addition to IP1) and notify CN about the availability of the new IP address through SCTP Address Dynamic Reconfiguration option [36]. This option defines two new chunk types (ASCONF and ASCONF-ACK) and several parameter types (Add IP Address, Delete IP address, and Set Primary Address, etc.).

STEP 3: Redirect data packets to new IP address

When MH moves further into the coverage area of wireless access network2, CN can redirect data traffic to new IP address IP2 to increase the possibility that data can be delivered successfully to the MH. This task can be accomplished by sending an ASCONF from MH to CN, through which CN sets its primary destination address to MH's IP2. At the same time, MH needs to modify its local routing table to make sure future outgoing packets to CN use the new path through AR2.

STEP 4: Update Location Manager (LM)

SIGMA supports location management by employing a location manager which maintains a database recording the correspondence between MH's identity and MH's current primary IP address. MH can use any unique information as its

22

identity, such as home address (like MIP), or domain name, or a public key defined in Public Key Infrastructure (PKI).

Following our example, once MH decides to handover, it should update the LM's relevant entry with the new IP address, IP2. The purpose of this procedure is to ensure that after MH moves from wireless access network1 into network2, subsequent new association setup requests can be routed to MH's new IP address (IP2). Note that his update has no impact on the existing active associations.

We can observe an important *difference* between SIGMA and MIP: the location management and data traffic forwarding functions are coupled together in MIP, while in SIGMA they are decoupled to speedup handover and make the deployment more flexible.

STEP 5: Delete or deactivate obsolete IP address

When MH moves out of the coverage of wireless access network1, no *new* or *retransmitted* data should be directed to address IP1. In SIGMA, MH notifies CN that IP1 is out of service for data transmission by sending an ASCONF chunk to CN to delete IP1 from CN's available destination IP list.

A less aggressive way to prevent CN from sending data to IP1 is to let MH advertise a zero receiver window (corresponding to IP1) to CN. This will give CN an impression that the interface buffer(to which IP1 is bound) is full and can not receive data any more. By deactivating, instead of deleting, the IP address, SIGMA can adapt more gracefully to MH's zigzag movement patterns and reuse the previous obtained IP address (IP1) as long as IP1's lifetime is not expired. This will reduce the latency and signalling traffic caused by obtaining a new IP address.

3.4   Timing Diagram

Fig. 3.6 summarizes the signalling sequences involved in SIGMA. The numbers before the events correspond to the step numbers in Sec. 3.3. Here we assume

IPv6 SAA is used by MH to get new IP address. It should be noted that before the old IP is deleted, it can receive data packets (not shown in the figure) in parallel with the exchange of signalling packets.

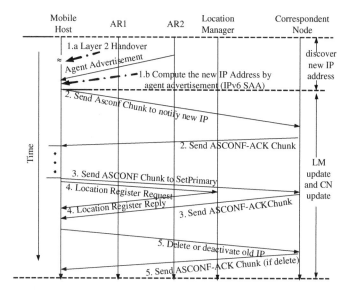

Figure 3.6: Timing diagram of SIGMA

3.5 Location Management

As mentioned in STEP 4 of Sec. 3.3, SIGMA needs to setup a location manager for maintaining a database of the correspondence between MH's identity and its current primary IP address. Unlike MIP, the location manager in SIGMA is not restricted to the same subnet as MH's home network (in fact, SIGMA has no concept of home or foreign network). The location of the LM does not have an impact on the handover performance of SIGMA. This will make the deployment of SIGMA much more flexible than MIP.

The location management can be done in the following sequence as shown in Fig. 3.7:

(1) MH updates the location manager with the current primary IP address.

(2) When CN wants to setup a new association with MH, CN sends a query to the location manager with MH's identity (home address, domain name, or public key, etc.)

(3) Location manager replies to CN with the current primary IP address of MH.

(4) CN sends an SCTP INIT chunk to MH's new primary IP address to setup the association.

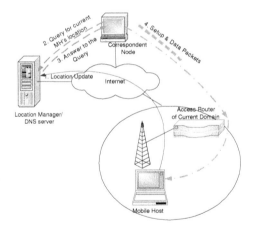

Figure 3.7: Location management in SIGMA

If we use the domain name as MH's identity, we can merge the location manager into a Domain Name Server (DNS). The idea of using DNS to locate mobile users can be traced back to Awerbuch et al. [37]. The advantage of this approach is its transparency to existing network applications that use domain name to IP address

25

mapping. An Internet administrative domain can allocate one or more location servers for its registered mobile users. Compared to MIP's requirement that each subnet must have a location management entity (HA), SIGMA can reduce system complexity and operating cost significantly by not having such a requirement. Moreover, the survivability of the whole system will also be enhanced as will be discussed in Chapter 6.

3.6   Summary

In this chapter, we have presented SIGMA, a Seamless IP-diversity-based Generalized Mobility Architecture, to manage handovers of mobile nodes. The handover process, timing diagram, and location management of SIGMA is described.

Chapter 4

Handover Performance Evaluation of SIGMA

The handover performance of SIGMA will be evaluated using two methods, simulation and analytical modeling. Sec. 4.1 describes the simulation topology and parameters. Sec. 4.2 presents the simulation results. Sec. 4.3 to 4.8 will discuss the analytical model that has been developed to evaluate SIGMA.

4.1 Simulation Topology and Parameters

This section describes the simulation topology and parameters that have been used to compare the performance of SIGMA and MIP. We have used the *ns-2* simulator [38] that supports SCTP as the transport protocol, and incorporated FMIPv6, HMIPv6, FHMIPv6 implementations [39] and MIP route optimization implementation [40]. We implemented SIGMA protocol in *ns-2* to support the simulation comparison (see Appendix B for details of the implementation).

4.1.1 Simulation Topology

The network topology used in our simulations for both MIPv6 and SIGMA is shown in Fig. 4.1. This topology has been used extensively in earlier MIP performance studies [4, 39]. In the figure, MIPv6 uses a HA, while SIGMA uses a Location Manager. Router2 in the topology acts as an Mobility Anchor Point (MAP) point in HMIPv6 and FHMIPv6, while it acts as only a normal router in FMIPv6 and

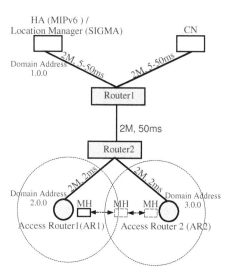

Figure 4.1: Simulation topology.

SIGMA. The link characteristics, namely the bandwidth (Megabits/s) and propagation delay (milliseconds), are shown on the links, which are similar to the values used in earlier simulation studies [39].

4.1.2 Simulation Parameters

Besides the simulation parameters listed in Table 4.1, we have also used the following configurations in our simulations:

- A pair of FTP source and sink agents are attached to the CN and MH, respectively, to transfer bulk data from CN to MH.

- Each Access Router (AR) has a radio coverage area of 40 meters in radius, and the overlapping region between two ARs is 10 meters. The advertisement period of the HA/AR1/AR2 is one second, but the advertisements from

Table 4.1: Simulation parameters for handover performance evaluation.

Wireless propagation model:	TwoRayGround
Antenna type:	Omni-Antenna
Antenna gain:	1.0
Topology size:	(200, 350)
Mac protocol	IEEE 802.11
Queue type:	Drop-tail
Queue size:	20 packets
SCTP *rwnd* limit	20 segments
Initial sender *cwnd*	2 segment
Initial sender *ssthresh*	20 segments
SCTP data chunk size:	512 bytes
Simulation time:	500 seconds

them are not synchronized. The radio coverage radius of 40 meters corresponds to the indoor IEEE 802.11 WLAN environment based on our testbed experiment and is also used by [39].

- To make a fair comparison, we have used standard SCTP protocol (without mobility related modifications) as the transport layer protocol for MIPv6 enhancements. This is to ensure that all the handover schemes use the same connection setup and congestion control control mechanisms, and that the results are only affected by the different handover schemes.

4.2 Simulation Results

This section shows the comparison results between SIGMA and MIPv6 enhancements in terms of handover latency, throughput, packet loss rate, and network friendliness.

4.2.1 Handover Latency

We define the *handover latency* as the time interval between the last data segment received through the old path and the first data segment received through the new

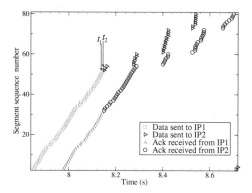

Figure 4.2: Segment sequence of SIGMA during one handover.

path from CN to MH. In this section, we will first show a packet trace of SIGMA to illustrate the seamless handover of SIGMA, then examine the impact of different parameters on the overall handover latency of SIGMA and MIPv6 enhancements.

4.2.1.1   Packet Trace of SIGMA

Fig. 4.2.1.1 shows the packet trace observed at the CN during one typical handover for SIGMA with data being sent from CN to MH. The segment sequence numbers are shown as modulo 100. From Fig. 4.2.1.1 we can observe that data segments are sent to MH's old IP address (2.0.1) until time 7.08 sec (point t_1), then the new IP address (3.0.1) almost immediately (point t_2), and all these packets are successfully delivered to MH. Therefore, SIGMA experienced a seamless handover because it could prepare the new path in parallel with data forwarding over the old path. This is the basic reason that explains why SIGMA can achieve a low handover latency, a low packet loss rate, and a high throughput as will be shown in the following comparison with MIPv6 enhancements.

30

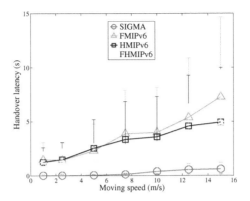

Figure 4.3: Impact of moving speed on SIGMA handover latency.

4.2.1.2 Impact of Moving Speed

We vary the moving speed of MH from 1.0m/s up to 15.0m/s. When MH moves faster, all MIPv6 enhancements and SIGMA will experience a higher handover latency due to shorter time to prepare for the handover (see Fig. 4.2.1.2). However, the increase in speed has the most significant effect on FMIPv6 since it relies on the assumption that detection of the new agent happened well in advance of the actual handover. When the moving speed is higher, the assumption can break down more easily. Because HMIPv6 and SIGMA do not rely on this assumption, the effect of moving speed is smaller. But when moving speed is higher, there is higher possibility that packets are forwarded to the outdated path and get lost; therefore the time instant that MH can receive packets from the new path will be postponed, and the handover latency increases accordingly. The 95% confidence interval for the handover latency results are also shown in Fig. 4.2.1.2.

4.2.1.3 Impact of Link Delay between HA (LM) and Router1

Next, we vary the link delay between HA(LM) and Router1 from 5ms up to 200ms. The link delay between HA(LM) and Router1 decides the time that it takes MH

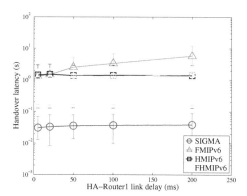

Figure 4.4: Impact of HA-Router1 delay on SIGMA handover latency.

to update the location registration, and the effect of this link delay on the overall latency is shown in Fig. 4.2.1.3 with 95% confidence interval. Since SIGMA decouples the location management function from the critical handover process (see STEP 4 of Fig. 3.6), this link delay does not have an impact on the latency of the SIGMA. This fact implies that we can put the location manager of SIGMA anywhere in the Internet without sacrificing handover performance. For HMIPv6 and FHMIPv6, when MH moves between AR1 and AR2, it only needs to register with the MAP node (Router2). Thus the link delay between HA and Router1 does not have much impact on these two enhancements of MIPv6. However, each location update in FMIPv6 needs to go through this link between HA and Router1, which will increases the overall latency with an increase of the link delay.

4.2.1.4 Impact of Link Delay between CN and Router1

Next, we vary the link delay between CN and Router1 from 5ms up to 200ms. The link delay between CN and Router1 decides the time that takes MH to update the binding cache at CN (or CN's protocol control block in OS kernel, in the case of SIGMA), the effect of this link delay on the overall latency is shown in

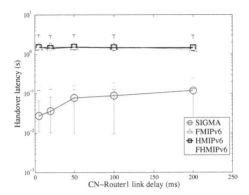

Figure 4.5: Impact of CN-Router1 delay on SIGMA handover latency.

Fig. 4.2.1.4 with 95% confidence interval. Our definition of handover latency does not require route optimization in versions of MIPv6 (binding update and return routability test) to finish. As long as the MH receives packets from the new path, either directly from CN or forwarded from HA, the handover is considered finished. Therefore, the link delay between CN and Router1 does not have much impact on the handover latency of MIPv6 enhancements. In contrast, SIGMA always requires updating CN before packets can be received from the new path. Therefore, the increase of this link delay will increase the handover latency (up to 109ms in the case of 200ms delay between CN and Router1).

4.2.2 Packet Loss Rate and Throughput

We define the *packet loss rate* as the number of lost packets due to handover divided by the total number of packets sent by CN. The *throughput* is defined as the total useful bits that can be delivered to MH's upper layer application divided by the simulation time, which gives us an estimate of average transmission speed that can be achieved. In this section, we will examine the impact of different parameters on

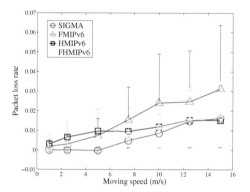

Figure 4.6: Impact of Moving Speed on packet loss rate.

the packet loss rate and throughput of SIGMA and MIPv6 enhancements. These parameters are the same ones as we have seen in Sec. 4.2.1.

4.2.2.1 Impact of Moving Speed

When MH moves faster, all versions of MIPv6 and SIGMA will experience a higher packet loss rate (Fig. 4.2.2.1) and decreased throughput (Fig. 4.2.2.1). This is because the possibility of packets being forwarded to the outdated path will increase with an increase in the speed. We can also notice that an increase in moving speed has the most significant effect on FMIPv6 since it relies on the assumption that detection of the new agent is well in advance of the actual handover, which may not hold when MH moves fast. Note that for SIGMA, an increase of moving speed from 12.5 to 15.0 results in a slight increase in throughput. This is because that rapid movement results in longer time for MH sitting still, i.e. less fraction of time spending during the movement. The 95% confidence interval for the packet loss rate and throughput results are also shown in Fig. 4.2.2.1 and 4.2.2.1.

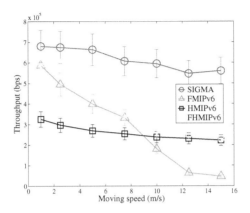

Figure 4.7: Impact of moving speed on throughput.

4.2.2.2 Impact of Link Delay between HA(LM) and Router1

As pointed out in Sec. 4.2.1.3, since SIGMA decouples the location management function from the critical handover process, this link delay does not have an impact on the packet loss rate and throughput of SIGMA (Figs. 4.2.2.2 and 4.2.2.2) with 95% confidence interval. For HMIPv6 and FHMIPv6, when MH moves between AR1 and AR2, it only needs to register with the MAP node (Router2), thus the link delay between HA and Router1 also does not have much impact on these two MIPv6 enhancements. However, each location update in FMIPv6 has to go through this link between HA and Router1, thus a higher delay in this link will result in packets forwarded by HA having an increased possibility of being sent to an outdated location and being dropped.

4.2.2.3 Impact of Link Delay between CN and Router1

As shown in Sec. 4.2.1.4, the link delay between CN and Router1 does not have an impact on the handover latency. As a result, the number of packets lost will remain the same with an increase of this link delay. However, a higher value of this link delay will increase RTT. Since the throughput of an SCTP association

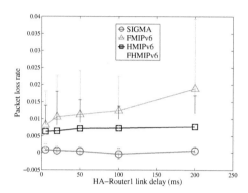

Figure 4.8: Impact of HA-Router1 delay on packet loss rate.

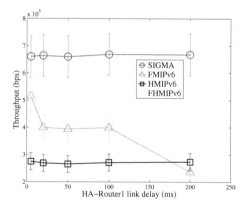

Figure 4.9: Impact of HA-Router1 delay on throughput.

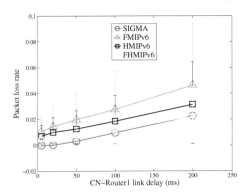

Figure 4.10: Impact of CN-Router1 delay on packet loss rate and throughput.

decreases as RTT increases, the total number of packets sent to the MH will decrease. When we compute packet loss rate by dividing the number of packets lost by the total number of packets sent by CN, the resulting loss percentage will increase (Figs. 4.2.2.3 and 4.2.2.3) with 95% confidence interval. For SIGMA, as this link delay increases, it has a negative effect on both packet loss (due to non-timely CN update) and throughput (longer RTT), so the packet loss rate increases relatively fast as compared to FHMIPv6 (Fig. 4.2.2.3).

4.2.3 Network Friendliness

A network friendly mobility protocol requires that when an MH enters a new domain, CN should probe for the new domain's network condition. In all MIP versions, CN's transport protocol stack is not aware of the handover, it continues to use the old congestion window (*cwnd*). As shown in Fig. 4.2.3, CN's *cwnd* remains constant after a handover around time 10.5 sec in the case where the handover latency is small enough that CN does not encounter a timeout resulting in drop of *cwnd*. This means that CN assumes the new network path to have the same capacity as the old one, which may cause network congestion if the new path does

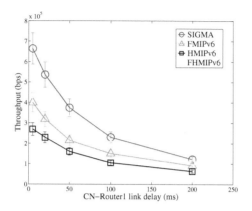

Figure 4.11: Impact of CN-Router1 delay on throughput.

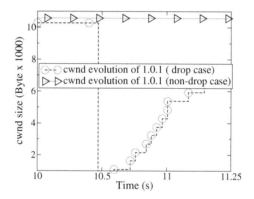

Figure 4.12: MIP *cwnd* evolution during handover.

not have enough capacity. Although this network unfriendliness can sometimes help MIP achieve better throughput, it is not preferable from the perspective of network performance. Note that in MIP, the sender may be forced to slow start after a handover, due to packet losses during handover, as shown in Fig. 4.2.3 where the CN goes through a slow start starting at around 10.6 secs. In contrast to MIP, SIGMA exhibits better network friendliness. The sender always probes

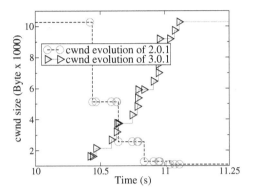

Figure 4.13: SIGMA *cwnd* evolution during handover.

the new network path after a handover, regardless of segment drops. As shown in Fig. 4.2.3, the new network path is used starting from time 10.4 sec when the CN automatically begins a slow start sequence to avoid any possible congestion. This is because the CN switches over to a new transport address, after a handover, which has different set of congestion control parameters from the old one.

Fig. 4.14 shows the CN's congestion window evolution within 100 seconds of simulation. The time instants labelled with odd subscripts (t_1, t_3, t_5, and t_7) stand for a handover happens from AP1 to AP2, while the ones labelled with even subscripts (t_2, t_4, t_6, and t_8) stand for a handover happens from AP2 to AP1. This figure shows that SIGMA can achieve seamless handover as evidenced by the fact that the *cwnd* for new path picks up before the *cwnd* for old path drops (which is due to no data being directed to the old path after new path becomes the primary path). Moreover the *cwnd* for new path is increased according to slow start algorithm to probe the new network gradually after each handover, which means SIGMA is network friendly.

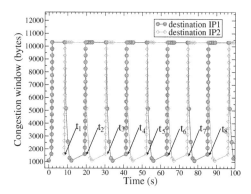

Figure 4.14: CN's congestion window in SIGMA during 100 second simulation time.

4.3   Analytical Model Overview

This section describes our overall modeling approach and the main structure of the analytical model. We consider the network topology shown in Fig. 4.15, which is a typical scenario for mobile handover. Here, the Correspondent Node (CN) is attached with N FTP flows which send data to Mobile Host (MH); AR1 and AR2 are two access routers, through which MH can connect into the network. (B_1, K_1) through (B_5, K_5) are the bandwidth and queue size of each corresponding link in the topology.

4.3.1   Overall Architecture

We model the throughput and packet loss of SIGMA using the fixed-point method [41]. Our overall model is split into two parts: source model and network model. The advantage of this methodology is its ability to isolate the analysis of SCTP's congestion control algorithms from network dynamics, rendering the model clear and accurate. The overall modeling architecture is shown in Fig.4.16. The output

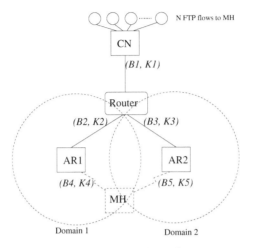

Figure 4.15: Network topology.

from the source model is fed into the network model as the arrival traffic and the output from the network model is fed back into the source model to compute the new arrival traffic pattern. This process is iterated until the subsequent iterations generate very similar results, which means the overall model has achieved an equilibrium point.

In the networking scenario shown in Fig. 4.15, packet losses may happen due to queue overflow at the link queues, wireless link corruption error, or mobile handovers. Data packets may also go through extra delay due to queuing, wireless media contention, or handover latency. According to the types of reasons that contribute to packet losses and delays, we further divide the Network Model in Fig. 4.16 into three sub-models: *Queue model*, *Wireless model*, and *Handover model*, which will be discussed in detail in Sec. 4.3.2. The traffic generated from Source model will be fed into the sub-models and the packet loss rate and delay obtained from separate sub-models will be combined and fed back to the Source model.

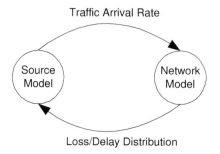

Figure 4.16: Overall modeling architecture.

4.3.2 Interaction between Source Model and Network Sub-models

The detailed feedback between Source Model and Network Sub-models is shown in Fig. 4.17. The function and inputs/outputs of each individual sub-model are described below:

- **Source model**: The Source model will capture the dynamics of SCTP congestion control.

  Inputs: Packet loss probability (combination of p_q, p_w, p_h) and packet delay (combination of d_q, d_w, d_h) output from the Queue model, Wireless model, and Handover model.

  Outputs: The number of SCTP sources (N) and the average arrival rate of individual SCTP sources (λ), which will be fed into Queue model, Wireless model, and Handover model.

- **Queue model**: The queue model will capture the packet loss and delay caused by queue waiting and overflow.

  Inputs: In addition to the traffic rate from the Source model, the input

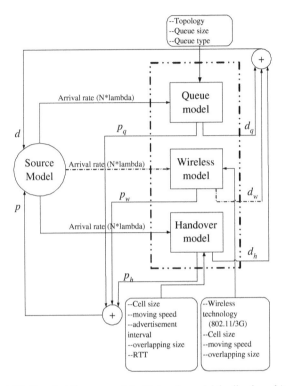

Figure 4.17: Detailed Source model - Network model feedback architecture.

includes network topology, queue size, service rate, and queue type.

Outputs: Packet loss probability and packet delay (p_q, d_q).

- **Wireless model**: The wireless model will capture wireless link corruption errors and packet losses due to user's mobility.

  Inputs: Other than output from source model, the input includes the wireless technology (802.11/ 3G), cell size, moving speed, overlapping size.

  Outputs: Packet loss probability and packet delay (p_w, d_w).

- **Handover model**: The handover model will capture the packet loss and delay in case the MH can't update CN fast enough after it receives advertisement from the new domain.

  Inputs: Other than traffic rate from the Source model, the input includes cell size, moving speed, advertisement interval, overlapping size, and RTT.

  Outputs: Packet loss probability and packet delay (p_h, d_h).

4.3.3   Convergence Criteria

After we obtain the value of packet loss probability p and delay d, they are fed back into the source model to compute the new generated traffic rate (λ_{source}). This traffic will then become the input traffic to the network model to recompute a new set of p and d. This process is iterated until the traffic rate (λ_{source}) generated from the previous iteration is close enough to the current iteration.

4.4   SCTP Source Model

In this section, we develop the average traffic rate generated by a SCTP source depending on an input of packet loss probability p and packet delay d. We first consider a single-homed SCTP association case then a multihomed association case. SCTP is based on the congestion control principles of TCP. Recently, several papers have reported analytical models to predict the throughput of TCP [41–44]. Since TCP does not support multihoming, the models did not consider the effect of multihoming on transport layer throughput, and thus cannot be readily applied for SCTP. *Our model differs from previous research in that the proposed model explicitly takes multihoming into account in the analysis.*

4.4.1 Modeling Assumptions

The assumptions we have made for developing the source model are described below, which are also used elsewhere [41, 45, 46].

- By considering a large number of SCTP sources' traffic as aggregated into the arrival traffic for the network, the overall traffic is regarded as a Poisson distribution for arrival;

- Loss between subsequent segments in the network are independent;

- Round Trip Time (RTT) has a exponential distribution;

- SCTP associations carry long-lived FTP traffic.

4.4.2 Notations for source model

The notations used in the source model are given below.

p_q, d_q Segment loss probability and mean delay obtained from the queueing network model, respectively.

d_{pt} Propagation and transmission delay between source and destination.

θ Round Trip Time (RTT) between source and destination; $\theta = d_{pt} + d_q$.

$cwnd$ Congestion window size (segments).

W_t Slow start threshold (segments).

$wmax$ Maximum value of $cwnd$.

N Number of SCTP sources.

T Value of Retransmission Time Out (RTO) (seconds).

$ccwnd, pcwnd$ Value of $cwnd$ size after and before a state transition.

π Steady state distribution of tuple $(cwnd, W_l, l)$.

$P_w(j)$ Probability of j segments lost in a window of size w.

P_w^{TO} Probability that a Time Out (TO) occurs when $cwnd = w$.

P_w^{FR} Probability that a Fast Retransmit (FR) occurs when $cwnd = w$.

$P\left(loss^{(k)}\right)$ Probability that k segments were lost during the last state transition.

$P\left(pcwnd^{(i)}, ccwnd^{(j)}\right)$ Probability that $pcwnd = i$ and $ccwnd = j$.

G Expected number of total segments generated by source model per RTT.

$E[L]$ Expected number of total losses per RTT.

λ_{source} Traffic rate generated by source model (segments/sec).

4.4.3 Single-homed SCTP Association

SCTP's congestion control is based on and very similar to the well proven rate-adaptive, window-based congestion control of TCP. The common features include the adoption of slow start, congestion avoidance, timeout and fast retransmit algorithms. However, there are several major differences between the congestion control mechanisms of TCP and SCTP. Since our modeling approach is based on that used for TCP [41], we list below the differences between the congestion control of TCP and SCTP.

- SCTP doesn't have an explicit fast-recovery phase. SCTP achieves fast recovery implicitly through the use of Selective Acknowledgment (SACK) [29].

- SCTP begins its slow start algorithm from $cwnd = 2$ instead of one as in TCP.

- The mandatory use of SACK in SCTP allows more robust reactions in the case of multiple losses from a single window of data. This avoids a time-consuming slow start stage after multiple segment losses, thus saving bandwidth and increasing throughput.

- TCP begins fast retransmit after the receipt of three Duplicate Acknowledgements (DupACKs); SCTP begins after four DupACKs. However, SCTP is able to clock out new data on receipt of the first three DupACKs, and can also retransmit a lost segment by ignoring whether the flight size is less than *cwnd*.

We show the state transition diagram of an SCTP association with one destination in Fig. 4.18; it is based on TCP's state transition diagram [41] and incorporates two differences between TCP and SCTP: (a) SCTP's slow start begins from two segments instead of one, (b) SCTP begins fast retransmit after four DupACKs, and therefore the triggering of fast retransmit in SCTP requires a current congestion window of at least five, whereas it is four for TCP.

In Fig. 4.18, every state includes three elements $(cwnd, W_t, l)$, where l is the loss indication: 0 means no loss occurred during previous transition and 1 means one or multiple losses occurred. For ease of reading, only $cwnd$ is shown in the circles, and thick circles correspond to states with $l = 1$. Here, $wmax = 16$ is assumed to model the largest receiver window ($rwnd$) of 16, and initial $W_t = wmax$. The rightmost column with thick circles denotes states undergoing fast retransmission. Since this column is identical for $W_t = 2, 4, 8, 16$, to keep the figure readable, only the case for $W_t = 2$ is shown.

The state transitions in Fig. 4.18 can be classified into four categories:

- ***Slow Start***: State transitions from $(w, W_t, 0)$ to $(2w, W_t, 0)$ with a transition rate of $P_w(0)/\theta$. This means the sender's congestion window size grows from w to $2w$ in one RTT, if there is no loss. For example, in Fig. 4.18, the transition probability from $cwnd = 4$ to 8 at $W_t = 16$ is $P_4(0)/\theta$.

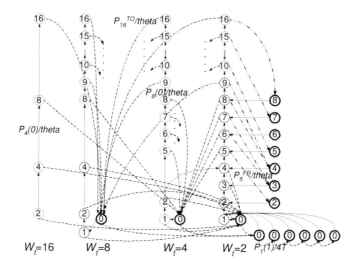

Figure 4.18: State transition of SCTP source - single-homed case.

- **Congestion Avoidance**: State transitions from $(w, W_t, 0)$ to $(w + 1, W_t, 0)$ with transition rate of $P_w(0)/\theta$. This means the sender's current window size grows from w to $w + 1$ in one RTT if there is no loss. For example, in Fig. 4.18, the transition probability from $cwnd = 8$ to 9 at $W_t = 4$ is $P_8(0)/\theta$.

- **Timeout**: State transitions from $(w, W_t, 0)$ to $(0, \lfloor w/2 \rfloor, 1)$ with transition rate of P_w^{TO}/θ. This means the sender's current window size drops from w to 0, and its slow start threshold drops from W_t to $\lfloor w/2 \rfloor$, and l changes from 0 to 1 within one RTT if timeout happens.

$$P_w^{TO} = \begin{cases} \sum_{i=1}^{w-4} P_w(i) \left(1 - (1 - p_q)^i\right) + \sum_{i=w-3}^{w} P_w(i) & \text{if } w \geq 5 \\ 1 - P_w(0) & \text{if } w < 5 \end{cases} \quad (4.1)$$

Although $cwnd = 1$ after a timeout in SCTP, we add the state $cwnd = 0$ as an intermediate state to model the waiting time before a timeout is detected. During this time, no segment is sent, so we count $cwnd$ as 0. For example, in Fig. 4.18, the transition probability from $(cwnd = 16, W_t = 4, 0)$ to $(cwnd = 0, W_t = 8, 1)$ is P_{16}^{TO}/θ.

- **_Exponential Backoff_**: State transitions from $(0, W_t, 1)$ to $(0, 2, 1)$ with transition rates of $P_1(1)/(2^j T)$, $j = 1, 2, \cdots, 6$ for jth successive timeout. In case of repeated timeouts, the SCTP sender will perform an exponential backoff. An example in Fig. 4.18 is the transition rate from the second to third timeout which is $P_1(1)/4T$.

- **_Fast Retransmit_**: state transitions from $(w, W_t, 0)$ to $(\lfloor w/2 \rfloor, \lfloor w/2 \rfloor, 1)$ with transition rate of P_w^{FR}/θ. This means that the sender's $cwnd$ drops from w to $\lfloor w/2 \rfloor$, the slow start threshold drops from W_t to $W_t/2$, and l changes from 0 to 1 in one RTT if a timeout happens.

$$P_w^{FR} = \begin{cases} 1 - P_w^{TO} - P_w(0) & \text{if } w \geq 5 \\ 0 & \text{if } w < 5 \end{cases} \tag{4.2}$$

For example, in Fig. 4.18, the transition rate from $(cwnd = 5, W_t = 2, 0)$ to $(cwnd = 2, W_t = 2, 1)$ is P_5^{FR}/θ.

If we assume packet losses to be independent from each other, $P_w(j)$ in Eqns. (4.1) and (4.2) can be determined by the Bernoulli formula: $P_w(j) = \binom{j}{w} p_q^j (1 - p_q)^{(w-j)}$.

After all transition rates in Fig. 4.18 are determined, the steady state distribution π of $(cwnd, W_t, l)$ can be calculated by:

$$\pi Q = \pi \tag{4.3}$$

where Q is the transition probability matrix.

4.4.4 Multihomed SCTP Association

We denote the expected number of segments generated by source model per RTT as:

$$G = \sum_{w=1}^{wmax} wP\left(cwnd^{(w)}\right) \tag{4.4}$$

By definition of π,

$$P\left(cwnd^{(w)}\right) = \sum_{W_t=2}^{wmax} \sum_{l=0}^{1} \pi(w, W_t, l) \tag{4.5}$$

To model an SCTP association with a multihomed destination, we next determine the traffic sent into the primary and alternative paths. We need to model SCTP's packet retransmission on the alternative path when there is a Time Out (TO) or a Fast Retransmit (FR). To do this, in Fig. 4.18, we strip the states where $l = 1$, and sum up all the losses when the system transits into these states (resulting from TO or FR) to obtain the total number of packets retransmitted on the alternative path, as shown in Fig. 4.19.

Bayes method is used to compute the expected number of segment losses during these types of transitions as described in detail in Sec. 4.4.5.

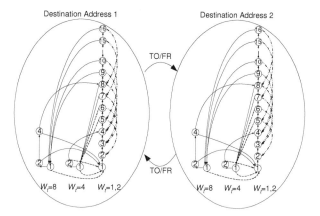

Figure 4.19: State transition of SCTP source - multihomed case.

4.4.5   Bayes Loss Estimation

We separate the reason for the transition to a state with $cwnd = w$ into two cases: due to a fast retransmit and due to a timeout. Then we combine these two cases to get the expected segment losses during the transition given $ccwnd = w$.

(1) Fast Retransmit case: Since $ccwnd = w$, the previous window size $pcwnd$ must be $2w$ or $2w + 1$. From Fig. 4.18, the $ccwnd$ can only range from 2 to $wmax/2$ after a Fast Retransmit. Moreover, the number of losses during this transition can not be more than $2w - 4$, otherwise a timeout will occur. From Bayes formula:

$$P\left(loss^{(k)}|pcwnd^{(i)}, ccwnd^{(w)}\right) = \frac{P\left(loss^{(k)}\right)P\left(pcwnd^{(i)}, ccwnd^{(w)}|loss^{(k)}\right)}{P\left(pcwnd^{(i)}, ccwnd^{(w)}\right)} \quad (4.6)$$

where $2 \leq w \leq wmax/2$, $i = 2w$ or $2w + 1$ and $1 \leq k \leq 2w - 4$.

Since we know that $P\left(loss^{(k)}\right) = P_i(k)$, and $P\left(pcwnd^{(i)}, ccwnd^{(w)}\right) = P_i^{FR}$, Eqn. (4.6) becomes:

$$P\left(loss^{(k)}|pcwnd^{(i)}, ccwnd^{(w)}\right) = \frac{P_i(k)P\left(pcwnd^{(i)}, ccwnd^{(w)}|loss^{(k)}\right)}{P_i^{FR}} \quad (4.7)$$

Next, we want to find $P\left(pcwnd^{(i)}, ccwnd^{(w)}|loss^{(k)}\right)$ in Eqn. (4.7). Since the transition to the current state has been due to a Fast Retransmit, given k segments lost from the original transmission, $ccwnd$ will become w only when all the successive retransmissions for the k segments are successful. A timeout will happen if any of the k retransmissions are lost. So, the conditional probability that $pcwnd$ was i and $ccwnd$ becomes w, given k losses happened, can be estimated as:

$$P\left(pcwnd^{(i)}, ccwnd^{(w)}|loss^{(k)}\right) = (1 - p)^k \quad (4.8)$$

By substituting Eqn. (4.8) into Eqn. (4.7), we can get:

$$P(loss^{(k)}|pcwnd^{(i)}, ccwnd^{(w)}) = \frac{P_i(k)(1-p)^k}{P_i^{FR}} \quad (4.9)$$

By summing up two cases for $i = 2w, 2w + 1$ in Eqn. (4.9), we can get the marginal conditional distribution:

$$P(loss^{(k)}|ccwnd^{(w)}) = \sum_{i=2w}^{2w+1} \frac{P_i(k)(1-p)^k}{P_i^{FR}} \quad (4.10)$$

(2) <u>Timeout</u> case: Here $ccwnd=0$ and $pcwnd$ could be any value from 1 to $wmax$, and $k = 1, 2, \ldots pcwnd$. Similarly, by Bayes Formula:

$$P\left(loss^{(k)}|pcwnd^{(w)}, ccwnd^{(0)}\right) = \frac{P(loss^{(k)})P(pcwnd^{(w)}, ccwnd^{(0)}|loss^{(k)})}{P(pcwnd^{(w)}, ccwnd^{(0)})} \quad (4.11)$$

Since we know that $P\left(loss^{(k)}\right) = P_w(k)$ and $P\left(pcwnd^{(w)}, ccwnd^{(0)}\right) = P_w^{TO}$, Eqn. (4.11) becomes:

$$P\left(loss^{(k)}|pcwnd^{(w)}, ccwnd^{(0)}\right) = \frac{P_w(k)P(pcwnd^{(w)}, ccwnd^{(0)}|loss^{(k)})}{P_w^{TO}} \quad (4.12)$$

Next, we want to find $P\left(pcwnd^{(w)}, ccwnd^{(0)}|loss^{(k)}\right)$ in Eqn. (4.12). Since the transition to the current state was caused by a timeout, given k segments were lost in the original transmission, if some of the retransmitted segments for the k segments failed or there are not enough DupACKs generated (in the case of $k = w - 3, w - 2, \ldots w$), $ccwnd$ will become zero; otherwise, a Fast Retransmit will happen. Also, because $pcwnd$ can be any value from 1 to $wmax$, we assume that $pcwnd$ ranges from 1 to $wmax$ with equal probability. So, the conditional probability that $pcwnd$ was w, $cwnd$ is 0, given k losses happen, can be estimated as:

$$P\left(pcwnd^{(w)}, ccwnd^{(0)}|loss^{(k)}\right) =$$
$$\begin{cases} \left[1 - (1-p)^k\right]/wmax & \text{for } k = 1, 2, \ldots w - 4 \\ 1/wmax & \text{for } k = w - 3, w - 2, \ldots w \end{cases} \quad (4.13)$$

Substituting Eqn. (4.13) into Eqn. (4.12), and summing up all the cases for $pcwnd = 1, 2, \ldots, wmax$, we get the marginal conditional distribution:

$$P\left(loss^{(k)}|ccwnd^{(0)}\right) = \sum_{pcwnd=1}^{wmax} P\left(loss^{(k)}|pcwnd^{(w)}, ccwnd^{(0)}\right)$$

$$= \begin{cases} \displaystyle\sum_{w=1}^{wmax} \frac{P_w(k)\left[1-(1-p)^k\right]}{wmax P_w^{TO}} \\ \qquad\qquad \text{for } k = 1, 2, \ldots w - 4 \\[2em] \displaystyle\sum_{w=1}^{wmax} \frac{P_w(k)}{wmax P_w^{TO}} \\ \qquad\qquad \text{for } k = w - 3, w - 2, \ldots w \end{cases} \qquad (4.14)$$

(3) <u>Combine FR and TO case</u>: Here $ccwnd = w$. This is done by weighting the number of segment losses (k) by the conditional probabilities (Eqns. (4.10) and (4.14)):

$$E\left[L|ccwnd = w\right] = \sum_{k=1}^{wmax} kP\left(loss^{(k)}|ccwnd^{(w)}\right) \qquad (4.15)$$

Finally, the overall expected segment losses occurring in the primary path, i.e., the traffic transferred into the alternative path, can be obtained using:

$$E[L] = \sum_{w=1}^{wmax} E\left[L|ccwnd = w\right] P(ccwnd = w)$$

$$= \sum_{w=1}^{wmax}\sum_{k=1}^{wmax} kP\left(loss^{(k)}|ccwnd^{(w)}\right) P(ccwnd = w) \qquad (4.16)$$

The above equation also represents the conditional expectation of segment losses occurring during transiting into all states with $l = 1$. We can thereby obtain the traffic on the primary path by subtracting the losses (which is also the traffic on the alternative path, Eqn. (4.16)) from the total traffic generated by the source (Eqn. (4.4)).

4.5 Queue Model

Solution of the source model in Sec. 4.4 requires the value of RTT ($\theta = d_{pt} + d_q$) and loss probability (p_q). In this section, we derive the values of d_q and p_q. In the network model, we consider two cases: the single queue case and the multi-queue case. In the single queue case, the whole network is modelled as an M/M/1/K queue [47]. In the multi-queue case, we consider all the queues in the network separately. We denote λ as the arrival traffic rate at a link queue (segments/sec), and μ, B, K as the service rate (segments/sec), bandwidth (bps), and buffer size (segments) of a link, respectively.

4.5.1 Single Queue Case

In Fig. 4.15, when B_2 through B_5 are large enough, the only queue that affects packet loss and delay is the SRC-Router queue. We can model the queuing network as an M/M/1/K queue with $K = K1$. We denote $\rho = \lambda/\mu$, where $\mu = B/8 * PacketSize$ (segments/sec). From M/M/1/K queuing theory [47], the segment loss probability can be calculated as:

$$p_q = \begin{cases} \frac{1}{K+1} & \rho \geq 1 \\ \frac{(1-\rho)\rho^K}{1-\rho^{(K+1)}} & \rho < 1 \end{cases} \tag{4.17}$$

To find the queuing delay (d_q), let S be the mean number of segments in the queue:

$$S = \begin{cases} \frac{K}{2} & \rho = 1 \\ \frac{\rho}{(1-\rho)} - \frac{K+1}{1-\rho^{(K+1)}}\rho^{K+1} & \rho \neq 1 \end{cases} \tag{4.18}$$

Considering the current segment being transmitted in the queue, we can obtain the mean queuing delay as:

$$d_q = \frac{S+1}{\mu} \tag{4.19}$$

4.5.2 Multi-queue Case

It has been shown previously [48] that in the presence of greedy connections (such as FTP) that tend to overload the network, different queueing models provide similar estimates of the average loss probability. Therefore, a simple queue for each link on the topology can be used to approximate the ensemble behavior the whole network. Other approaches with significantly greater complexity (mainly based on group arrivals and services) were also tested by previous authors [48], but results do not change significantly in the case of long-lived flows. In Fig. 4.15, if (B_2, K_2) through (B_5, K_5) are finite, we assume that the queuing network can be modelled as a combination of M/M/1/K queues, as shown in Fig. 4.20. The input

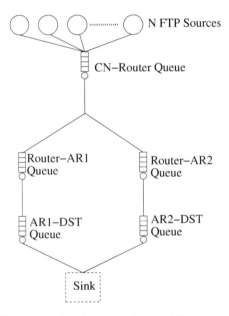

Figure 4.20: Queuing network for multi-queue case.

traffic into each queue in Fig. 4.20 can be determined as: $\lambda_{SRC-Router} = \lambda_{source}$.

This means the input traffic to the SRC-Router queue is the same as the traffic generated from the source model. We can also get the input traffic to the Router-AR1 queue as:

$$\lambda_{Router-AR1} = \lambda_{SRC-Router}(1-R)(1-p_{SRC-Router}) \qquad (4.20)$$

where $p_{SRC-Router}$ denotes the loss probability at the SRC-Router queue which can be determined using Eqn. (4.17) with $\lambda = \lambda_{SRC-Router}$, $B = B1$, and $K = K1$. R is the percentage of packets retransmitted through the alternative path (via AR2), which can be determined as:

$$R = E(L)/G \qquad (4.21)$$

where $E(L)$ is the expected number of packet losses during one RTT i.e. those that will be retransmitted through the alternative path, and G (determined by Eqn. (4.4)) is the total traffic generated by the source model. Similarly, we can get the input traffic to the AR1-MH queue:

$$\lambda_{AR1-DST} = \lambda_{Router-AR1}(1-p_{Router-AR1}) \qquad (4.22)$$

Since each queue is modelled as an M/M/1/K queue, we can use Eqns. (4.17) and (4.18) to get loss probability and average queue occupancy of each individual queue. Assuming no repeated losses for traffic retransmitted into the alternative path, we can get the overall loss probability at the primary path as:

$$p_q = 1 - (1-p_{SRC-Router})(1-p_{Router-AR1})(1-p_{AR1-DST}) \qquad (4.23)$$

where $p_{Router-AR1}$ and $p_{AR1-DST}$ denote the loss probability at Router-AR1 queue and AR1-DST queue, respectively. This means that the overall loss probability is the percentage of packets that did not successfully go through all the three queues.

By Little's law, we can model the average delay in the queuing network as:

$$d_q \;=\; \frac{S}{\lambda}$$

$$= \frac{S_{SRC-Router} + S_{Router-AR1} + S_{AR1-DST}}{\lambda} \tag{4.24}$$

where $S_{SRC-Router}$, $S_{Router-AR1}$, and $S_{AR1-DST}$ denote the average queue occupancy at SRC-Router queue, Router-AR1 queue, and AR1-DST queue, respectively; λ is the input traffic rate at the SRC-Router queue. Similarly, by substituting loss probability and queue occupancy of the queues in alternative path into Eqns. (4.23) and (4.24), we can get p_q and d_q for the alternative path.

4.6 Wireless Model

Over the past 30 years, many wireless propagation models have been proposed for wireless link budget design. Several of them are more frequently used: *Free-space*, *Two-ray ground*, and *Log-normal shadowing* models [8]. The Free-space model and the Two-ray ground model predict the received power as a deterministic function of distance. In reality, the received power at certain distance is a random variable due to the effect of environment shadowing, which may cause differences in the received power at two different locations having the same transmitter-receiver distance. So the more general and widely-applicable model is the Log-normal shadowing model or shadowing model for short, which will also be used in this chapter.

The shadowing model depicts that at any given point with distance of d between transmitter and receiver, the received power at that point can be calculated by [8]:

$$P_r(d)[dBm] = P_t[dBm] - PL(d)[dB] \tag{4.25}$$

where $P_r(d)$ is the received power at distance d from transmitter and P_t is the transmitter power. $PL(d)$ is called path loss at distance d, which in turn can be calculated by:

$$PL(d)[dB] = \overline{PL(d)} + X_\delta = \overline{PL(d_0)} + 10n\log\left(\frac{d}{d_0}\right) + X_\delta \tag{4.26}$$

where X_δ is a zero-mean Gaussian distributed random variable in dB with standard deviation δ (also in dB); d_0 is a reference point where there exists a line-of-sight path to the transmitter and the received signal strength can be precisely measured; n is called the *path loss exponent*, which is normally ranging from 4 to 6 for obstructed indoor environments; and δ can be computed from measured data, while 4 is commonly used for simulation and analysis [8].

The shadowing model can be used to determine the probability that received signal strength is smaller than a given receiving threshold. Let γ be the receiving threshold, then:

$$P[P_r(d) < \gamma] = Q(\frac{\overline{P_r(d)} - \gamma}{\delta})$$ (4.27)

where Q function is defined as:

$$Q(z) = \frac{1}{\sqrt{2\pi}} \int_z^\infty \exp\left(-\frac{x^2}{2}\right) dx$$

Also the area coverage percentage based on a given receiving threshold can be computed. For a circular coverage area having radius R, the percentage of area with a received signal that is larger than threshold γ can be computed as:

$$U(\gamma, R) = \frac{1}{\pi R^2} \int_0^{2\pi} \int_0^R P[P_r(d) > \gamma] r \, dr \, d\theta$$ (4.28)

It has been shown [8] that $U(\gamma, R)$ can be simplified to:

$$U(\gamma, R) = \frac{1}{2} \left(1 - \mathrm{erf}(a) + \exp\left(\frac{1 - 2ab}{b^2}\right)\left[1 - \mathrm{erf}\left(\frac{1 - ab}{b}\right)\right]\right)$$ (4.29)

where

$$a = \frac{\left(\gamma - \overline{P_r(d_0)} + 10n\log(R/d_0)\right)}{\sqrt{2}\delta}$$

$$b = \frac{10n\log e}{\sqrt{2}\delta}$$

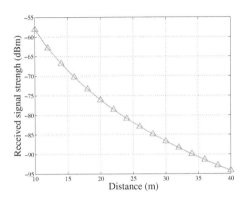

Figure 4.21: Received signal strength as a function of distance.

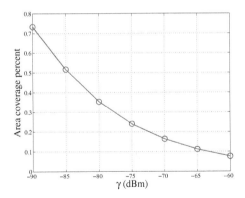

Figure 4.22: Area coverage percentage as a function of receiving threshold.

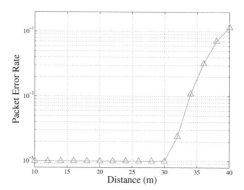

Figure 4.23: Packet error rate as a function of distance.

We performed some testing in our testbed to characterize the packet error rate generated by the shadowing model. We choose $d_0 = 5$m and measured $\overline{Pr(d_0)} \approx 40$dBm. With $n = 6$ and $\delta = 4$, we can plot $\overline{Pr(d)}$ as shown in Fig. 4.21. We can see from the figure at a distance of 30 meters the received signal drops below -85dBm, which will begin to produce a high packet error rate based on our measurements.

Next, we choose $R = 40$ and $\gamma = [-90, -60]$ dBm and plot the area coverage percentage based on Eqn. 4.29. We can see that a higher γ means that it requires higher received power at the receiver to correctly detect the signal, therefore, a lower percentage of the area can satisfy this power requirement.

For a packet being transmitted through a wireless link, it generally also needs to go through modulation, spectrum spreading, and channel coding procedures to reduce the packet error rate. If we choose IEEE 802.11b *DBPSK* modulation, *barker sequence*-based direct sequence spectrum spreading and assume the channel coding procedures can catch up to two bits of error in one packet, then we can plot the resulting packet error rate as a function of distance from transmitter as shown in Fig. 4.23. It can be observed in Fig. 4.23 that for our testbed and model settings mentioned above, within 30 meters of distance the packet error rate is very small,

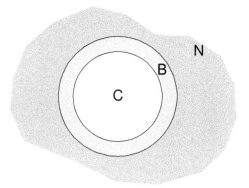

Figure 4.24: Three types of radio coverage.

while for distances between 30 to 40 meters the packet error rate ramps up very fast.

For ease of using the wireless model, we can just classify the radio coverage of a particular AP into three categories: Center coverage (**C**), Border coverage (**B**) and No coverage (**N**) as shown in Fig. 4.24. For any given wireless cells, we can first determine d_0, $P_r(d_0)$, γ, n, δ, then use the model to find out the boundaries between coverage categories and associated packet error rate.

4.7   Handover Model

In this section, the handover model is developed to capture the packet loss rate resulting from SIGMA handover. First, Sec. 4.7.1 presents the notations used in the handover model. Then, Sec. 4.7.2 describes the actual model which is based on Continuous Time Markov Chains (CTMC) [47, 49].

4.7.1　Notations for Handover Model

The notations used in the handover model are listed below:

D Overlapping distance.

T_{L2} Layer 2 connection setup latency.

T_{IAR} IP address resolution latency.

RTT Round trip time between MH and CN.

RTO Retransmission timer value.

P_{wb} Wireless channel loss rate when MH is in border coverage.

P_q Queueing network loss rate.

P_h Handover loss rate.

T_r Subnet Residence time.

v MH moving speed.

4.7.2　CTMC Model for SIGMA Handover

A CTMC is used to characterize the state transition of MH. For the sake of simplicity, we assume the transition times have exponential distribution. The states in Fig. 4.25 are defined as follows:

HI Handover initiated.

AS Add_IP sent over old path.

SS Set_Primary sent over old path.

HC Handover Complete.

ARX1-3 First, second, and third ADD_IP retransmission.

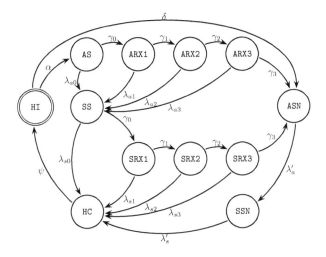

Figure 4.25: CTMC model for SIGMA handover.

SRX1-3 First, second, and third Set_PRIMARY retransmission.

ASN Add_IP sent over new path.

SSN Set_Primary sent over new path

The transition rates in Fig. 4.25 are computed based on parameters like Layer 2 setup latency, IP address resolution latency, RTT, wireless channel error and queuing error, RTO, overlapping distance, moving speed, and subnet residence time. The state transition in Fig. 4.25 works as follows:

- The handover preparation begins when MH receives Layer 2 beacons, which is represented by HI state.

- After MH finishes Layer 2 connection setup, it sends ADD_IP chunk to CN and enters into AS state. The transition rate between HI and AS can be computed as: $\alpha = 1/(T_{L2} + T_{IAR})$.

63

- This ADD_IP chunk could be dropped since MH has moved close to the boundary of the old cell, or the losses in the queuing network could also drop the ADD_IP chunk. If this happens, the MH would wait until timeout and retransmit the ADD_IP chunk and enters into `ARX1` state. If consecutive drops occur, MH will move into `ARX2` or even `ARX3` states. The transition rates for timeout events can be computed as: $\gamma_i = (P_q + P_{wb} - P_q P_{wb})/RTO$, where $i = 1, 2, 3$ and $(P_q + P_{wb} - P_q P_{wb})$ means a packet drop happened either in the wireless channel or the queuing network. We assume the timer values stay constant here, and other retransmitting algorithms can be used such as binary exponential backoff.

- If the ADD_IP is delivered successfully, MH would receive a ACK after one RTT, then MH will transmit SET_PRIMARY when it moves beyond the central point of overlapping distance and enters into `SS` states. The transition rate between `AS` and `SS` can be computed as: $\lambda_{a0} = (1 - P_q)(1 - P_{wb})/\max(\text{RTT}, \frac{D}{2v})$, where $(1 - P_q)(1 - P_{wb})$ means no packet drop happened in wireless channel and queuing network.

- The success of retransmission of ADD_IP will also make MH enter into the `SS` state. The transition rate between `AS` and `SS` can be computed as: $\lambda_{aj} = (1 - P_q)(1 - P_{wb})/\max(RTT, \frac{D}{2v} - jRTO), j = 1, 2, 3$.

- When SET_PRIMARY is delivered successfully, MH would receive a ACK after one RTT, then MH will enters into `HC` state. Therefore, the transition rate is $\lambda_{s0} = (1 - P_q)(1 - P_{wb})/RTT$.

- Similar to the case of transmission ADD_IP, the loss of SET_PRIMARY will make MH enter into `SRX1`, `SRX2` and `SRX3` states, and the retransmission success will move MH from `SRX1`, `SRX2` and `SRX3` states to `HC` state. The transition rates are: $\lambda_{sj} = (1 - P_q)(1 - P_{wb})/RTT, j = 1, 2, 3$.

- If ADD_IP or SET_PRIMARY is not delivered after three retransmissions, this indicates MH has most probably moved out of the overlapping region. MH will move into `ASN` state and begin send ADD_IP through the new path.

- Since MH has entered into the new cell when it is in `ASN` state, in order to simplify the Markov Chain, we assume there is no loss between the transition of `ASN`, `SSN` and `HC` states. Therefore, the transition rates are: $\lambda'_a = \lambda'_s = 1/RTT$.

- MH enters into `HC` states, which means it has completed this instance of handover. After T_r time, a new handover begins, thus the MH moves into `HI` state again. Therefore, the transition rate between `HC` and `HI` states are $\psi = 1/Tr$.

- If MH can not finish all the signaling before it moves out of the overlapping region, it will enter into `ASN` state (transition rate: $\delta = v/D$), and begin sending ADD_IP through the new path.

In order to compute the stationary distribution of states in Fig. 4.25, the states are numbered in the following order starting from one: {1: `HI`, 2: `AS`, 3: `ARX1`, 4: `ARX2`, 5: `ARX3`, 6: `SS`, 7: `ASN`, 8: `SRX1`, 9: `SRX2`, 10: `SRX3`, 11: `SSN`, 12: `HC`}. Then we can write the infinitesimal generator matrix $\boldsymbol{Q}$ of CTMC as follows:

$$
\begin{pmatrix}
-\alpha - \delta & \alpha & & & & & \delta & & & & & \\
 & -\gamma_0 - \lambda_{a0} & \gamma_0 & & & \lambda_{a0} & & & & & & \\
 & & -\gamma_1 - \lambda_{a1} & \gamma_1 & & \lambda_{a1} & & & \boldsymbol{0} & & & \\
 & & & -\gamma_2 - \lambda_{a2} & \gamma_2 & \lambda_{a2} & & & & & & \\
 & & & & -\gamma_3 - \lambda_{a3} & \lambda_{a3} & \gamma_3 & & & & & \\
 & & & & & -\gamma_0 - \lambda_{s0} & \gamma_0 & & & & & \lambda_{s0} \\
 & & & & & & -\lambda'_a & & & & & \lambda'_a \\
 & & & \boldsymbol{0} & & & & -\gamma_1 - \lambda_{s1} & \gamma_1 & & & \lambda_{s1} \\
 & & & & & & & & -\gamma_2 - \lambda_{s2} & \gamma_2 & & \lambda_{s2} \\
 & & & & & & & \gamma_3 & & -\gamma_3 - \lambda_{s3} & & \lambda_{s3} \\
 & & & & & & & & & & -\lambda'_s & \lambda'_s \\
-\psi & & & & & & & & & & & \psi
\end{pmatrix}
$$

Once we have determined the infinitesimal generator matrix $\boldsymbol{Q}$, we can compute the stationary distribution of the CTMC $\boldsymbol{\pi}$ by:

$$\boldsymbol{\pi}\boldsymbol{Q} = \boldsymbol{0} \qquad (4.30)$$

The packet loss probability due to handover will be the sum of steady state probability of all states other than HI, AS, SS, and HC states. Therefore, the packet loss probability can be calculated by:

$$P_h = \boldsymbol{\pi}\boldsymbol{S}^T \qquad (4.31)$$

where $\boldsymbol{S} = [0, 0, 1, 1, 1, 0, 1, 1, 1, 1, 1, 0]$.

4.8 Numerical Results

In this section, the source model and three network sub-models: queue model, wireless model, and handover model are combined together according to the feedback structure presented in Sec. 4.3.2. The generated numerical results are presented below. The performance measures used in this section are handover packet loss rate and average end-to-end throughput.

4.8.1 Impact of RTT and Layer 2 Connection Setup Latency

The impact of RTT and T_{L2} on handover packet loss rate is shown in Fig. 4.26. We can see that with the increase of RTT or T_{L2}, the packet loss rate resulting from handover increases. This is because an increase of RTT means a longer time to update CN about the current location of MH, so there is a higher possibility that packets are delivered to an outdated location. Also, a higher Layer 2 connection setup latency will postpone the beginning of SIGMA handover, and thus produce a higher risk of not finishing handover before MH moves out of the overlapping region, which in turn will result in higher packet loss probability.

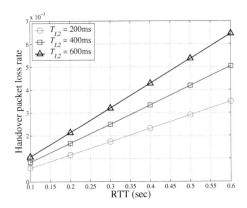

Figure 4.26: Impact of RTT and T_{L2} on handover packet loss rate.

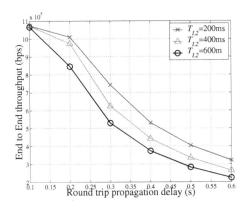

Figure 4.27: Impact of RTT and T_{L2} on end-to-end throughput.

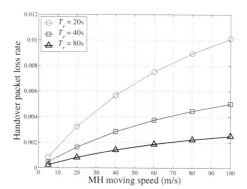

Figure 4.28: Impact of moving speed and residence time on handover packet loss rate.

Fig. 4.27 shows the impact of RTT and T_{L2} on end-to-end throughput. The explanation above for the packet loss rate also explains the trend of end-to-end throughput. Basically, a higher packet loss rate will produce a lower end-to-end throughput, since the sender will trigger the congestion control algorithm when the packet loss is detected. Also, for window based transport protocols like SCTP, a higher RTT will limit the rate of pumping data into the network, which will further reduce the end-to-end throughput.

4.8.2 Impact of Moving Speed and Subnet Residence Time

The impact of moving speed and subnet residence time on handover packet loss rate is shown in Fig. 4.28. We can see that with the increase of moving speed, the packet loss rate resulting from handover increases. This is because MH will have less time to prepare for the handover when moving speed is higher. If MH can not receive the ACK for SET_PRIMARY before it moves out of the overlapping region, the packets are delivered to the old location, which will cause packet losses up to one window. With the increase of subnet residence time (T_r), the packet

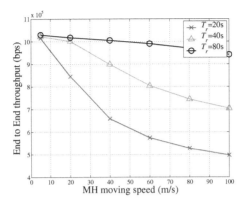

Figure 4.29: Impact of moving speed and residence time on end-to-end throughput.

loss rate is lower. This is because MH will perform handover less frequently with higher residence time, and on average the long-term packet loss rate decreases accordingly.

4.8.3 Impact of Overlapping Distance and IP Address Resolution Latency

The impact of overlapping distance and IP address resolution latency on handover packet loss rate is shown in Fig. 4.30. We can see that with the increase of overlapping distance, the packet loss rate resulting from handover decreases. This is because a longer overlapping distance means MH has more time to perform SIGMA signaling before it moves out of the old subnet, therefore, the packet has a higher possibility of being forwarded to the correct location when MH changes point of attachment. The impact of IP address resolution latency is similar to that of Layer 2 connection setup latency. The increase of IP address resolution latency postpones the start of SIGMA handover and increases the packet loss probability.

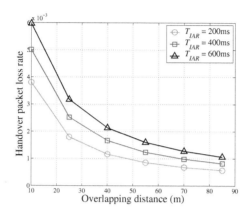

Figure 4.30: Impact of overlapping distance and T_{IAR} on handover packet loss rate.

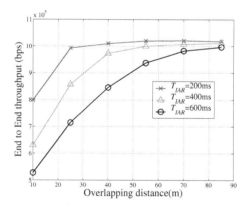

Figure 4.31: Impact of overlapping distance and T_{IAR} on end-to-end throughput.

4.9 Summary

In this chapter, we evaluated the handover performance of SIGMA through simulation and compared with that of MIPv6 enhancements. Different performance measures, including handover latency, packet loss and throughput, have been compared.

We also developed an analytical model for SIGMA, which consists of a source model, queueing sub-model, wireless propagation sub-model, and handover sub-model. Through simulation and numerical results, we have shown that SIGMA can achieve a seamless handover with low handover latency, low packet loss rate and high end-to-end throughput. SIGMA has also been shown to be network *friendly* by probing the new network at every handover.

Chapter 5

Signaling Cost Analysis of SIGMA

SIGMA relies on the signaling message exchange between the MH, correspondent node (CN), and location manager (LM). For every handover, MH need to send binding update and location update to CN and LM, respectively. For SIGMA to be useful in real world wireless system, all these signaling messages should not cost too much network bandwidth to leave no space for payload data transmission.

The signaling cost analysis for MIP protocols were presented earlier in [50, 51], but there is no work done in analyzing the signaling cost of transport layer mobility solutions. The *objective* of this section is to look into the signaling cost required by SIGMA.

The rest of this chapter is structured as follows: the analytical model for SIGMA signaling cost is presented in Sec. 5.1 and Sec. 5.2. Then we evaluate the signaling cost of SIGMA by the model under various input parameters in Sec. 5.5.

5.1   Modeling Preparation

In this section, we describe some necessary preparation work for developing an analytical model for SIGMA signaling cost. First, the network structure we are considering and model's assumptions and notations are presented in Secs. 5.1.1, 5.1.2 and 5.1.3 respectively. Then the MH mobility model and traffic arrival model used by signaling cost analysis are set up in Secs. 5.1.4 and 5.1.5 respectively. After these modeling foundations are ready, Sec. 5.2 develops the signaling cost for location update, binding update and packet delivery in SIGMA.

5.1.1 Network Structure

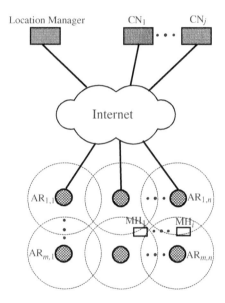

Figure 5.1: Network structure considered.

In this section, we describe the network structure that will be used in our analytical model, which is shown in Fig. 5.1. In the figure, a two dimensional subnet arrangement is assumed for modeling MH movement. $AR_{1,1}, \cdots AR_{m,n}$ stand for the access routers. There are one location manager and a number of CNs connected into the topology by Internet. The MHs are roaming around in the subnets covered by $AR_{1,1}, \cdots AR_{m,n}$, and each of them are communicating with one or more of the CNs. Between a pair of MH and CN, intermittent file transfers occur caused by mobile user request information from CNs using protocols like HTTP. We call each active transferring period during the whole MH-CN interactivity as one session.

5.1.2 Model Assumptions

The assumptions we have made for developing our analytical model of SIGMA signaling cost are described below.

- In the previous study of P-MIP signaling cost analytical model [51], the session time is assumed to be Pareto distribution and the session arrival is assumed to be *poisson* distribution. In our modeling process, Both session time and session interval time are of *Pareto* distribution to better model HTTP traffic [52,53], which is dominant in current Internet traffic load. The *Pareto* distribution is a heavy-tailed distribution, and it can be characterized with two parameters: minimum possible value (κ), and a heavy-tailness factor (σ).

- Mobile host moves according to Random Waypoint model [54], which is the most frequently used model in recent mobile networking research. In this mobility model, a MH randomly selects a destination point in the topology area according to uniform distribution, then moves towards this point at a random speed again uniformly selected between (v_{min}, v_{max}). This one movement is called an *epoch*, and the elapsed time and the moved distance during an epoch are called *epoch time* and *epoch length*, respectively. At destination point, the MH will stay stationary for a period of time, called *pause time*, after that a new epoch starts.

- Processing costs at the endpoints (MH and CN) are not counted into the total signaling cost since these costs stand for the load that can be scattered into user terminals. Because we are more concerned about the load on the network elements, this assumption enables us to concentrate on the impact of protocol on the network performance. This same assumption was also made by other previous works [50, 51, 55].

5.1.3 Notations

The notations used in this paper are given below.

l_{ml} average distance between MH and location manager in hops.

l_{mc} average distance between MH and CN in hops.

N_{mh} total number of MHs.

N_{cn} average number of CNs with which a MH is communicating.

LU_{ml} transmission cost of a location update from MH to location manager.

γ_l processing cost at location manager for each location update.

v_l location database lookup cost per second for each transport layer association at LM.

Ψ_{LU} location update cost per second for the whole system, including transmission cost and processing cost incurred by location update of all MHs, $\Psi_{LU} = N_{mh}\frac{LU_{ml}+\gamma_l}{T_r}$.

BU_{mc} transmission cost of a binding update between MH and CN.

Ψ_{BU} binding update cost per second between MHs and CNs for the whole system, $\Psi_{BU} = N_{mh}N_{cn}\frac{BU_{mc}}{T_r}$.

Ψ_{PD} packet delivery cost per second from CNs to MHs for the whole system.

Ψ_{TOT} total signaling cost per second for the whole system including location update cost, binding update cost and packet delivery cost, $\Psi_{TOT} = \Psi_{LU} + \Psi_{BU} + \Psi_{PD}$.

D_{pq} average propagation and queuing delay per hop.

$E(T)$ expected value of *epoch time*.

$E(P)$ expected value of MH pause time between movements.

$E(L)$ expected value of *epoch length*.

$E(C)$ expected number of subnet crossings per *epoch*.

v moving speed of MH.

T_r MH residence time in a subnet.

T_s session time.

T_i session interval time.

κ_s minimum session time.

σ_s heavy-tailness factor for session time.

BW_{mc} bottleneck bandwidth between CN and MH.

κ_i minimum session interval time.

σ_i heavy-tailness factor for session interval time.

λ_a average session arrival rate.

5.1.4   Mobility Model

The objective of this section is to find the average residence time (T_r) for MH in a subnet. With this parameter, we know the frequency for MH to change the point of attachment, therefore the frequency of updating LM and CN. T_r can be estimated by the time between two successive movements (*epoch time* plus *pause time*) divided by the number of subnet crossing during this epoch, as shown in Eqn. (5.1):

$$T_r = \frac{E(T) + E(P)}{E(C)} \qquad (5.1)$$

We first compute $E(T)$, since *epoch length* L and movement speed v are independent:

$$E(T) = E(L/v) = E(L)E(1/v) \qquad (5.2)$$

Since the moving speed is of uniform distribution between (v_{min}, v_{max}), we have:

$$
\begin{aligned}
E(1/v) &= \int_{v_{min}}^{v_{max}} (1/v) \frac{1}{v_{max} - v_{min}} dv \\
&= \frac{ln(v_{max}/v_{min})}{v_{max} - v_{min}}
\end{aligned}
\qquad (5.3)
$$

In order to determine $E(L)$ and $E(C)$, the shape of subnets and their arrangement pattern need to be fixed down first. It is very difficult to have a general model that can handle every kind of possible shapes and arrangement patterns. For the sake of tractability, we assume an arrangement of circular subnets in a rectangular topology as shown in Fig. 5.2, and m, n are the number of vertically and horizontally arranged subnets in the topology, respectively. This assumption is an ideal abstraction of real-world wireless networks and serves as a start point for our mobility model. More practical shapes and patterns which applies to specific scenarios can be used when they are available.

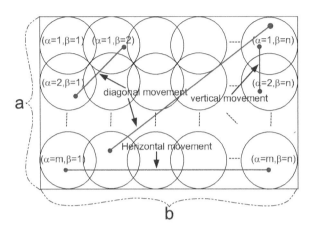

Figure 5.2: Arrangement of subnets in a rectangular topology

From [54], we know that $E(L)$ for a rectangular area of size $a \times b$ can be estimated as:

$$
\begin{aligned}
E(L) &= \frac{1}{15}\left[\frac{a^3}{b^2} + \frac{b^3}{a^2} + \sqrt{a^2 + b^2}\left(3 - \frac{a^2}{b^2} - \frac{b^2}{a^2}\right)\right] \\
&+ \frac{1}{6}\left[\frac{b^2}{a}\Phi\left(\frac{\sqrt{a^2 + b^2}}{b}\right) + \frac{a^2}{b}\Phi\left(\frac{\sqrt{a^2 + b^2}}{a}\right)\right] \\
&\text{where } \Phi(\cdot) = \ln\left(\cdot + \sqrt{(\cdot)^2 - 1}\right).
\end{aligned}
\tag{5.4}
$$

Now we can get $E(T)$ by combining Eqns. (5.2), (5.3) and (5.4). Since pause time has been assumed to be uniform distribution between $(0, P_{max})$, we have:

$$
E(P) = \int_0^{P_{max}} \frac{P}{P_{max}} dP = P_{max}/2
\tag{5.5}
$$

78

Next, we need to find $E(C)$, the general form of which can be expressed as [54]:

$$E(C) = \frac{1}{m^2 n^2} \sum_{\alpha_j=1}^{m} \sum_{\beta_j=1}^{n} \sum_{\alpha_i=1}^{m} \sum_{\beta_i=1}^{n} C\begin{pmatrix} (\alpha_i, \beta_i) \\ (\alpha_j, \beta_j) \end{pmatrix} \tag{5.6}$$

The value $C\begin{pmatrix} (\alpha_i, \beta_i) \\ (\alpha_j, \beta_j) \end{pmatrix}$ is the number of subnet crossing caused by a movement between subnet (α_i, β_i) to (α_j, β_j), which depends on the actual subnet shape and arrangement. Consider the circular subnet arrangement as shown in Fig. 5.2, we can observe three kind of movements: horizontal, vertical and diagonal. $C\begin{pmatrix} (\alpha_i, \beta_i) \\ (\alpha_j, \beta_j) \end{pmatrix}$ can be generalized by the following Manhattan distance metric:

$$C\begin{pmatrix} (\alpha_i, \beta_i) \\ (\alpha_j, \beta_j) \end{pmatrix} = |\alpha_i - \alpha_j| + |\beta_i - \beta_j| \tag{5.7}$$

By substituting Eqn. (5.7) into Eqn. (5.6), we can get the expression for $E(C)$:

$$E(C) = \frac{1}{m^2 n^2} \sum_{\alpha_j=1}^{m} \sum_{\beta_j=1}^{n} \sum_{\alpha_i=1}^{m} \sum_{\beta_i=1}^{n} (|\alpha_i - \alpha_j| + |\beta_i - \beta_j|) \tag{5.8}$$

Substituting Eqns. (5.2), (5.5) and (5.8) into Eqn. (5.1), we can get the expression for T_r.

5.1.5 Arrival Traffic Model

The objective of this section is to find the average session arrival rate (λ_a). As discussed in Sec. 5.1.2, both session time and session interval time are of *Pareto* distribution. The PDF function of session time's distribution is [52]:

$$f_{T_s}(t) = \frac{\sigma_s \kappa_s^{\sigma_s}}{t^{(\sigma_s+1)}} \tag{5.9}$$

where $\sigma_s = 1.2$, and κ_s can be estimated as:

$$\kappa_s = \frac{10\text{KB}}{BW_{mc}} + l_{mc}D_{pq} \tag{5.10}$$

Also from [52], we know session interval time has a PDF function of:

$$f_{T_i}(t) = \frac{\sigma_i \kappa_i^{\sigma_i}}{t^{(\sigma_i+1)}} \tag{5.11}$$

where $\sigma_i = 1.5$, and $\kappa_i = 30s$.

Consider k $(k > 0)$ consecutive user session arrivals (the start of the session $k+1$ means the end of the session k plus an interval time) as shown in Fig. 5.3, the total time for k sessions can be calculated as:

$$T_{tot} = k(T_s + T_i) \tag{5.12}$$

So, the session arrival rate is:

$$\lambda_a = \frac{k}{E(T_{tot})} = \frac{1}{E(T_s) + E(T_i)} \tag{5.13}$$

From probability theory, since $T_s > 1$ and $T_i > 1$, the expected value of T_s and T_i are:

$$E(T_s) = \int_0^\infty t f_{T_s}(t)dt = \frac{\kappa_s \sigma_s}{\sigma_s - 1} \tag{5.14}$$

$$E(T_i) = \int_0^\infty t f_{T_i}(t)dt = \frac{\kappa_i \sigma_i}{\sigma_i - 1} \tag{5.15}$$

By substituting Eqns. (5.14) and (5.15) into Eqn. (5.13), we can get the average session arrival rate.

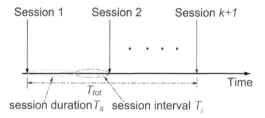

Figure 5.3: Session arrival illustration.

5.2 Signaling Cost Analysis of SIGMA

In this section, the signaling cost of SIGMA will be analyzed. Subsections 5.2.1, 5.2.2, and 5.2.3 develop the cost for location update, binding update and packet delivery, respectively. Finally, subsection 5.2.4 gives the total signaling cost of SIGMA.

5.2.1 Location Update Cost

In SIGMA, every subnet crossing (happens every T_r seconds) by an MH will trigger a location update, which incurs a transmission cost (LU_{ml}) and processing cost (γ) for the location update message. Since there is only one location update per subnet crossing, no matter how many CNs an MH is communicating with, the number of CNs does not have any impact on the location update cost. Therefore, the average location update cost per second in the whole system can be estimated as the number of MHs multiplied by the location update cost for each MH, divided by the average subnet residence time:

$$\Psi_{LU}^T = N_{mh}\frac{LU_{ml} + \gamma_l}{T_r} \tag{5.16}$$

Due to frame retransmissions and medium access contentions at the data link layer of wireless links, transmission cost of a wireless hop is higher than that of a wired hop; we denote this effect by a proportionality constant, ρ. Let the per-hop location update transmission cost be δ_U, for a round trip, LU_{ml} can be calculated as:

$$LU_{ml} = 2(l_{ml} - 1 + \rho)\delta_U \qquad (5.17)$$

Where $(l_{ml} - 1)$ represents the number of wired hops. Therefore,

$$\Psi_{LU}^T = N_{mh}\frac{2(l_{ml} - 1 + \rho)\delta_U + \gamma_l}{T_r} \qquad (5.18)$$

5.2.2 Binding Update Cost

In the analysis of binding update cost, processing costs at the endpoints (MH and CN) are not counted into the total signaling cost, since these costs stand for the load that can be scattered into user terminals and hence do not contribute to the network load. Because we are more concerned about the load on the network elements, this assumption enables us to concentrate on the impact of the handover protocol on network performance. This same assumption was also made by other previous works [50, 51, 55].

Similar to the analysis in Sec. 5.2.1, every subnet crossing will trigger a binding update to CN, which incurs a transmission cost (BU_{mc}) due to the binding update message. For each CN communicating with an MH, the MH need to send a binding update after each handover. Therefore, the average binding update cost can be estimated as:

$$\Psi_{BU}^T = N_{mh}N_{cn}\frac{BU_{mc}}{T_r} \qquad (5.19)$$

Let the per-hop binding update transmission cost be δ_B. The BU_{mc} can be calculated as:

$$BU_{mc} = 2(l_{mc} - 1 + \rho)\delta_B \qquad (5.20)$$

Therefore, the binding update cost per second in the whole system can be calculated by multiplying the number of MHs, the average number of communicating CNs, and the average cost per binding update:

$$\Psi_{BU}^T = N_{mh} N_{cn} \frac{2(l_{mc} - 1 + \rho)\delta_B}{T_r} \tag{5.21}$$

5.2.3 Packet Delivery Cost

Unlike the analysis of packet delivery cost in [50], we do not consider the data packet transmission cost, IP routing table searching cost, and bandwidth allocation cost since these costs are incurred by standard IP switching, which are not particularly related to mobility protocols. Instead, we only consider the location database lookup cost at LM. Moreover we take into account the processing cost caused by packet tunnelling to better reflect the impact of mobility protocol on overall network load.

For SIGMA, a location database lookup at LM is required when an association is being setup between CN and MH. If each session duration time is independent from each other, the association setup event happens every S/λ_{sa} seconds. If we assume the database lookup cost has a linear relationship with N_{mh}, and φ_l and ψ be the per location database lookup cost and the linear coefficient at LM, then the per-second per-association lookup cost v_l can be calculated as:

$$v_l = \frac{\varphi_l \lambda_{sa}}{S} = \frac{\psi N_{mh} \lambda_{sa}}{S} \tag{5.22}$$

Since SIGMA is free of packet encapsulation or decapsulation, there is no processing cost incurred at intermediate routers. So the packet delivery cost from CN to MH can be calculated by only counting the location database lookup cost. This cost can be expressed as:

$$\Psi_{PD}^T = N_{mh} N_{cn} v_l$$

83

$$= N_{mh}^2 N_{cn} \frac{\psi \lambda_{sa}}{S} \tag{5.23}$$

5.2.4 Total Signaling Cost of SIGMA

Based on above analysis on the location update cost, binding update cost, and packet delivery cost shown in Eqns. (5.18), (5.21), and (5.23), we can get the total signaling cost of SIGMA as:

$$\Psi_{TOT}^T = \Psi_{LU}^T + \Psi_{BU}^T + \Psi_{PD}^T \tag{5.24}$$

5.3 Review of Hierarchical Mobile IPv6

In this section, we also compare the signaling cost of SIGMA with HMIPv6. We choose HMIPv6 as the benchmark protocol for signaling cost comparison because HMIPv6 is designed to reduce the signaling cost of base MIPv6, and it has the lowest signaling cost in all versions of MIPv6 enhancements. We, therefore, briefly describe the HMIPv6 first in this section.

The objective of HMIPv6 is to reduce the frequency and delay of location updates caused by MH's mobility. In HMIPv6, operation of the correspondent node and HA are the same as MIPv6. A new network element, called the Mobility Anchor Point (MAP), is used to introduce hierarchy in mobility management. A MAP covers several subnets under its domain, called a *region* in this paper. A MAP is essentially a local Home Agent. The introduction of MAP can limit the amount of MIPv6 signalling cost outside its region as follows:

- When an MH roams between the subnets within a region (covered by a MAP), it only sends location updates to the local MAP rather than the HA (that is typically further away and has a higher load).

- The HA is updated only when the MH moves out of the region.

HMIPv6 operates as follows. An MH entering a MAP domain receives Router Advertisements containing information on one or more local MAPs. The MH updates the HA with an address assigned by the MAP, called Regional COA (RCoA), as its current location. The MAP intercepts all packets sent to the MH, encapsulates, and forwards them to the MH's current address. If the MH changes its point of attachment within a MAP domain, it gets a new local CoA (LCoA) from the AR serving it; the MH only needs to register the LCoA with the MAP. MH's mobility (change of the LCoA) is transparent to the HA, and the RCoA remains unchanged (thus no need to update HA) as long as the MH stays within a MAP's region.

5.4 Signaling Cost Analysis of HMIPv6

The analysis in this section follow a logic which is similar to the previous work on HMIP signaling cost analysis [50]. However, *our analysis differs from [50] in three ways*: (i) we do not consider the packet delivery costs incurred by standard IP switching, since they are not particularly related to mobility protocols; (ii) the tunnelling costs at HA and MAP are considered explicitly; (iii) we removed the processing costs at FAs to match the operation of HMIPv6. These modifications to the analysis of [50] enables us to compare the signaling cost of SIGMA and HMIPv6 more consistently. In HMIPv6, there is no binding update cost since the MH will not send a binding update to CN (if we consider HMIPv6 operating at the *bidirectional tunnelling mode* [2]). Secs. 5.4.1 and 5.4.2 develop the cost for location update and packet delivery respectively, and Sec. 5.4.3 gives the total signaling cost of HMIPv6.

5.4.1 Location Update Cost

In HMIPv6, an MH does not need to register with the HA until the MH moves out of the region covered by a MAP, instead it only registers with the MAP.

Therefore, every subnet crossing within a MAP (happens every T_r seconds) will trigger a registration to the MAP, which incurs a transmission cost to MAP (LU_{mm}) and processing cost at MAP (γ_m) of the location update message. Therefore, $C_{mm} = LU_{mm} + \gamma_m$.

For every *region* crossing between MAPs (happens every $M \times T_r$ seconds), MH needs to register with HA, which incurs a transmission cost to HA (LU_{mh}), processing cost at HA (γ_h), and processing cost at MAP ($2\gamma_m$, since MAP needs to process both registration request and reply messages). Therefore, $C_{mh} = LU_{mh} + \gamma_h + 2\gamma_m$.

Similar to SIGMA, the number of CNs that an MH is communicating with have no impact on the location update. Therefore, the average location update cost per second in the whole system can be estimated as the number of MHs multiplied by the location update cost for each MH, then divided by the average subnet residence time:

$$\Psi_{LU}^{H} = N_{mh} \frac{MC_{mm} + C_{mh}}{MT_r} \tag{5.25}$$

Similar to Eqn. (5.17), for a round trip, LU_{mh} and LU_{mm} can be calculated as:

$$LU_{mh} = 2(l_{mm} + l_{mh} - 1 + \rho)\delta_U \tag{5.26}$$

$$LU_{mm} = 2(l_{mm} - 1 + \rho)\delta_U \tag{5.27}$$

Also, M can be calculated from the total number of subnets ($m \times n$) and the number of subnets beneath a MAP (R): [50]:

$$M = 1 + \frac{mn - 1}{mn - R} \tag{5.28}$$

Therefore,

$$\Psi_{LU}^{H} = N_{mh} \left[\frac{2(l_{mm} - 1 + \rho)\delta_U + \gamma_m}{T_r} + \frac{2(l_{mm} + l_{mh} - 1 + \rho)\delta_U + \gamma_h + 2\gamma_m}{T_r} \times \frac{mn - R}{2mn - R - 1} \right] \tag{5.29}$$

5.4.2 Packet Delivery Cost

Similar to the analysis of Sec. 5.2.3, for packet delivery cost analysis, we only consider the location database lookup cost and tunnelling-related costs at HA and MAP. For each packet sent from CN to MH, processing costs incurred in sequence are: one location database lookup and one encapsulation at HA; one location database lookup, one decapsulation and one encapsulation at MAP.

Let φ_h, φ_m be the per location database lookup costs at HA, MAP, respectively; let τ be the per encapsulation/decapsulation cost at HA or MAP; and let ψ be the linear constant for location database lookup as defined in Eqn. (5.22); then we have:

$$v_h = \varphi_h + \tau = (\psi N_{mh}) + \tau \tag{5.30}$$

$$v_m = \varphi_m + 2\tau = \left(\psi \frac{N_{mh}R}{mn}\right) + 2\tau \tag{5.31}$$

So the packet delivery cost from CN to MH can be calculated by summing up the processing cost due to database lookup and tunnelling in the system, as shown in Eqns. (5.30) and (5.31). This cost can be expressed as:

$$\begin{aligned}
\Psi_{PD}^{H} &= N_{mh}N_{cn}\lambda_{pa}(v_h + v_m) \\
&= N_{mh}N_{cn}\lambda_{pa}\left(\psi N_{mh}\frac{mn+R}{mn} + 3\tau\right)
\end{aligned} \tag{5.32}$$

Where packet arrival rate (λ_{pa}) can be calculated from the session arrival rate and packet size. Let F be the file size being transferred by the session, and $PMTU$ be the path MTU between CN and MH, then the packet arrival rate can be calculated as:

$$\lambda_{pa} = \lambda_{sa}\frac{F}{PMTU} \tag{5.33}$$

5.4.3 Total HMIPv6 Signaling Cost

Based on above analysis of the location update cost and packet delivery cost shown in Eqns. (5.29) and (5.32), we can get the total signaling cost of HMIPv6 as:

$$\Psi_{TOT}^{H} = \Psi_{LU}^{H} + \Psi_{PD}^{H} \tag{5.34}$$

5.5 Results and Signalling Cost Comparison of SIGMA and HMIPv6

In this section, we present results showing the effect of various input parameters on SIGMA's total signaling cost. In all the numerical examples, using the following parameter values, which are obtained from previous work [50] and our calculation based on user traffic and mobility models [52,54]: $\gamma_l = 30$, $\psi = 0.3$, $F = 10$Kbytes, $PMTU = 576$bytes, $S = 10$, $\rho = 10$, $l_{ml} = 35$, $l_{mc} = 35$, $m = 10$, $n = 8$, $R = 10$, $\gamma_h = 30$, $\gamma_m = 20$, $\tau = 0.5$, $\lambda_{sa} = 0.01$, $l_{mh} = 25$, and $l_{mm} = 10$.

5.5.1 Impact of Number of MHs under Different Maximum MH Moving Speeds

The impact of number of MHs on total signaling cost of SIGMA and HMIPv6 for different MH moving speed is shown in Figs. 5.4 and 5.5. Here, the values used for other parameters are: $N_{cn} = 1$ and $\delta_U = \delta_B = 0.2$. From the figures, we can see that under different moving speeds, the signaling cost of both SIGMA and HMIPv6 increases with the increase of the number of MHs.

When the moving speed is higher, the subnet residence time T_r decreases (see Eqns. (5.1), (5.2), and (5.3)), resulting in a increase of the location update and binding update costs per second (see Eqns. (5.18) and (5.21)). We can also observe that the total signaling cost of SIGMA is less than HMIPv6 in this scenario; this is because when δ_U and δ_B are small, the location update and binding update costs

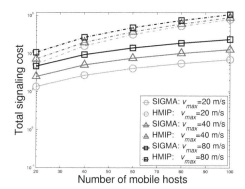

Figure 5.4: Impact of number of MHs on total signaling cost of SIGMA and HMIPv6.

are not high, and the high packet delivery cost will make the signaling cost of HMIPv6 much higher than that of SIGMA. Fig. 5.5 confirms that even for a very high moving speed of MH (300m/s), SIGMA still has a lower signaling cost that HMIPv6.

5.5.2 Impact of Average Number of Communicating CN and Location Update Transmission Cost

Next, we set subnet residence time $T_r = 60s$, and number of MHs $N_{mh} = 80$. The impact of the number of average CNs with which an MH communicates with for different per-hop transmission cost for location update cost (δ_U) is shown in Fig. 5.6. It can be observed from this figure that when the average number of communicating CNs increases, the total signaling cost increases (see Eqns.(5.18), (5.21) and (5.23)). Also, when δ_U increases, the location update cost per second will increase as indicated by Eqn. (5.17), which will result in the increase of the total signaling cost of both SIGMA and HMIPv6. However, we can see that the impact of δ_U is much

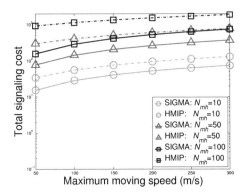

Figure 5.5: Impact of moving speed on total signaling cost of SIGMA and HMIPv6.

smaller in HMIPv6; this is because HMIPv6's signaling cost is less sensitive to location update cost due to its hierarchical structure. In this scenario, signaling cost of HMIPv6 is higher than that of SIGMA when $\delta_U = 0.4$ or 1.6. However, when $\delta_U = 6.4$, SIGMA requires a higher signaling cost due to frequent location update for each subnet crossing (compared to HMIPv6's hierarchical mobility management policy).

5.5.3 Session to Mobility Ratio

Session to Mobility Ratio (SMR) is a mobile packet network's counterpart of Call to Mobility Ratio (CMR) in PCS networks. We vary T_r from 75 to 375 seconds with λ_{sa} fixed to 0.01, which yields a SMR of 0.75 to 3.75. The impact of SMR on total signaling cost for different N_{mh} is shown in Fig. 5.7. We can observe that a higher SMR results in lower signaling cost in both SIGMA and HMIPv6. This is mainly because high SMR means lower mobility, and thus lower signaling cost due to less location update and binding update. Also, we can see that the decrease of HMIPv6's signaling cost as a function of SMR is not as fast as that of SIGMA.

90

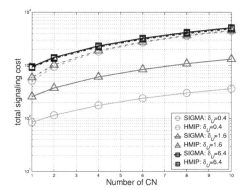

Figure 5.6: Impact of number of CNs and per-hop binding update transmission cost

This again is because HMIPv6's hierarchy structure reduces the impact of mobility on the signaling cost. The signaling cost, therefore, decreases slower than that of SIGMA when MH's mobility decreases.

5.5.3.1 Relative Signaling Cost of SIGMA to HMIPv6

Fig. 5.8 shows the impact of (location update transmission cost) / (packet tunnelling cost) ratio (δ_U/τ) on the relative signaling cost between SIGMA and HMIPv6. A higher δ_U/τ ratio means that the location update requires more cost while packet encapsulation/decapsulation costs less. This ratio depends on the implementation of the intermediate routers. We can see that as long as $\delta_U/\tau < 15$, the signaling cost of SIGMA is less than that of HMIPv6 due to the advantage of no tunnelling required. After that equilibrium point, the cost of location update will take dominance, and the signaling cost of SIGMA will become higher than that of HMIPv6.

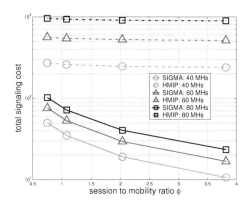

Figure 5.7: Impact of *SMR* on total signaling cost for different N_{mh}

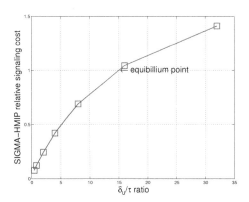

Figure 5.8: Impact of δ_U/τ ratio on SIGMA to HMIPv6 relative signaling cost

5.6 Summary

In this chapter, we have evaluated the signaling cost of SIGMA and compared with that of HMIPv6 using an analytical model. Numerical results show that, in most scenarios, the signaling cost of SIGMA is lower than HMIPv6. However, there is a tradeoff between location update transmission cost (δ_U) and packet tunnelling cost (τ); very high δ_U/τ ratio results in the signaling cost of SIGMA being higher than that of HMIPv6.

Chapter 6

Survivability Analysis of SIGMA

When using MIP in a mobile computing environment, one of the concern is low survivability due to its single-point failure of Home Agents. Mobile IP is based on the concept of Home Agent (HA) for recording the current location of the Mobile Host (MH) and forwarding packets to MH when it moves out of its home network. In MIP, the location database of all the mobile nodes are distributed across all the HAs that are scattered at different locations (home networks). According to principles of distributed computing, this approach appears to have good survivability. However, there are two major drawbacks to this location management scheme as given below:

- Each user's location and account information can only be accessible through its HA. The transparent replication of the HA, if not impossible, is not an easy task as it involves extra signaling support as proposed in [56].

- HAs have to be located in the home network of an MH in order to intercept the packets sent to the MH. The complete home network could be located in a hostile environment, in the case of failure of the home networks, all the MHs belonging to the home network would no longer be accessible.

The location management and data traffic forwarding functions in SIGMA are decoupled, allowing it to overcome the drawbacks of MIP in terms of survivability. In SIGMA, Location Managers (LM) can be combined with DNS servers, which can be deployed anywhere in the Internet and in a highly secure location. Also, it would be fairly straightforward to duplicate the LMs since they are not responsible for user data forwarding.

In the literature, two recent papers that have addressed the problem of MIP survivability are [57] and [58]. Ref [57] proposed a procedure to let MH register with multiple MAPs to avoid single point failure. Ref [58] used a similar idea as SIGMA, and the authors proposed a way to move HA (they call it Location Register) to a secure location and duplicate HA through some translation servers or a Quorum Consensus algorithm borrowed from distributed database systems. But none of the papers analytically models the survivability of MIP. Through analytical models, the *objective* of this paper is to show that the location management scheme used in SIGMA can enhance the survivability of the mobile network. The *contributions* of the current chapter can be summarized as:

- Illustrate the reason of SIGMA can achieve better survivability than MIP.

- Develop a analytical model based Markov Reward Process to determine the survivability of location management schemes.

- Compare the survivability of SIGMA and MIP in terms of system availability and user response time.

The rest of this chapter is structured as follows: Sec. 3.5 reviews the location management scheme used by SIGMA, Sec. 6.1 illustrates the basic reason of SIGMA being able to achieve better survivability than MIP. The analytical model is described in Sec. 6.2 and the numerical results are shown in Sec. 6.3.

6.1 Survivability Comparison of SIGMA and MIP

In this section we discuss the survivability of MIP and SIGMA. We highlight the disadvantages of MIP in terms of survivability, and then discuss how those issues are taken care of in SIGMA.

6.1.1 Survivability of MIP

In MIP, the location database of all the mobile nodes are distributed across all the HAs that are scattered at different locations (home networks). According to principles of distributed computing, this approach appears to have good survivability. However, there are two major drawbacks to this distributed nature of location management as given below:

- If we examine the actual distribution of the mobile users' location information in the system, we can see that each user's location and account information can only be accessible through its HA; these information are not truly distributed to increase the survivability of the system. The transparent replication of the HA, if not impossible, is not an easy task as it involves extra signaling support as proposed in [56].

- Even if we replicate HA to another agent, these HAs have to be located in the home network of an MH in order to intercept the packets sent to the MH. The complete home network could be located in a hostile environment, such as a battlefield, where the possibility of all HAs being destroyed is still relatively high. In the case of failure of the home networks, all the MHs belonging to the home network would no longer be accessible.

6.1.2 Centralized Location Management of SIGMA Offers Higher Survivability

Referring to Fig. 3.7, SIGMA uses a centralized location management approach. As discussed in Sec. 3.5, the location management and data traffic forwarding functions in SIGMA are decoupled, allowing it to overcome many of the drawbacks of MIP in terms of survivability (see Sec. 6.1.1) as given below:

- The LM uses a structure which is similar to a DNS server, or can be directly combined with a DNS server. It is, therefore, easy to replicate the Location Manager of SIGMA at distributed secure locations to improve survivability.

- Only location updates/queries need to be directed to the LM. Data traffic do not need to be intercepted and forwarded by the LM to the MH. Thus, the LM does not have to be located in a specific network to intercept data packets destined to a particular MH. It is possible to avoid physically locating the LM in a hostile environment; it can be located in a secure environment, making it highly available in the network.

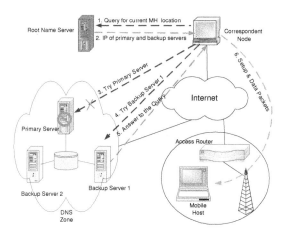

Figure 6.1: Survivability of SIGMA's location management.

Fig. 6.1 illustrates the survivability of SIGMA's location management, implemented using DNS servers as location servers. Currently, there are 13 servers in the Internet [59] which constitute the root of the DNS name space hierarchy. There are also several delegated name servers in the DNS zone [18], one of which is primary and the others are for backup and they share a common location database. If an

MH's domain name belongs to this DNS zone, the MH is managed by the name servers in that zone. When the CN wishes to establish a connection with the MH, it first sends a request to one of the root name servers, which will direct the CN to query the intermediate name servers in the hierarchy. At last, CN obtains the IP addresses of the name servers in the DNS zone to which the MH belongs. The CN then tries to contact the primary name server to obtain MH's current location. If the primary server is down, CN drops the previous request and retries backup name server 1, and so on. When a backup server replies with the MH's current location, the CN sends a connection setup message to MH. There is an important difference between the concept of MH's DNS zone in SIGMA and MH's home network in MIP. The former is a logical or soft boundary defined by domain names while the latter is a hard boundary determined by IP routing infrastructure.

If special software is installed in the primary/backup name servers to constitute a high-availability cluster, the location lookup latency can be further reduced. During normal operation, heart beat signals are exchanged within the cluster. When the primary name server goes down, a backup name server automatically takes over the IP address of the primary server. A query requests from a CN is thus transparently routed to the backup server without any need for retransmission of the request from the CN.

Other benefits SIGMA's centralized location management over MIP's location management can be summarized as follows:

- *Security*: Storing user location information in a central secure database is much more secure than being scattered over various Home Agents located at different sub-networks (in the case of Mobile IP).

- *Scalability*: Location servers do not intervene with data forwarding task, which helps in adapting to the growth in the number of mobile users gracefully.

98

- *Manageability*: Centralized location management provides a mechanism for an organization/service provider to control user accesses from a single server.

6.2   Analytical Model

The aim of our model is to perform a combined analysis of system availability and performance evaluation. J. Meyer created a new measure called *performability* in [60,61], which will be used in this paper to measure the survivability of a system. A performability model consists of a availability sub-model, a performance sub-model, and a glue model that combine these two sub-models. We choose *Markov Reward Model* as the glue model since it provides a natural framework for an integrated specification of state transitions due to server failures and the system performance (equivalent to reward) under each system state.

6.2.1   Networking Architecture

The networking architecture been considered in the analytical model is shown in Fig. 6.2. The router in Fig. 6.2 forwards location updates from MHs, location queries from CNs, and Distributed Denial of Service (DDoS) attack traffic [62] to N location managers according to a round-robin policy. Each location manager has an independent queue of size K packets. After being processed by one of location managers, the acknowledgement/reply to the update/query/attack packets are transmitted back to their originators.

6.2.2   Assumptions and Notations

We have made the following assumptions in our analytical model to make it computationally tractable:

- Arrival of location updates, queries, and DDoS attacks are *Poisson* processes.

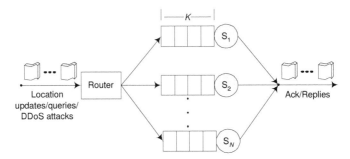

Figure 6.2: Queuing model of N location managers

- Location managers can not differentiate DDoS attack traffic from legitimate traffic.

- All location managers share common set of MH's mobility bindings.

- Processing time of location updates, queries, and DDoS attacks are exponential distributed and have same mean value.

- Hardware failures can be perfectly covered[1], i.e. system can degrade gracefully when one of the working server fails.

- Hardware failures always occurs on the servers with heaviest load.

Following are the notations that will be used in the analytical model:

N total number of location managers.

λ_u, λ_q, λ_a arrival rate of location updates, queries, and DDoS attack, respectively.

λ summation of λ_u, λ_q, λ_a.

---

[1]In an imperfect coverage system, some failures are impossible to be detected and the failure of one component will halt the whole system.

μ location manager processing rate.

K queue size of each location manager (packets).

γ, δ hardware failure rate and repair rate, respectively.

τ mean time to failure (MTTF)

ϕ mean time to repair (MTTR)

6.2.3 Combined System Availability & Performance Model for SIGMA Survivability

The objective of our model is to determine the average response time and blocking probability of SIGMA under the impact of hardware failures and DDoS attacks. We use a two-dimensional Continuous Time Markov Chains (CTMC) to capture system characteristics. The state transition diagram is shown in Fig. 6.3, in which each state is labeled as (N_w, L), where N_w is the number of currently working servers and L is the total number of packets in the system. When N_w equals N, since each server has a queue size of K, the maximum value of L is $K'' = N \times K$. Similarly, When N_w equals $N - 1$, the maximum value of L is $K' = (N - 1) \times K$.

We illustrate the transition diagram through several examples:

- current state is $(N,0)$, the hardware failure of any one server (happens with a rate of $N\gamma$) will make the next state $(N - 1,0)$.

- current state is $(N,1)$, arrival of one update/query/attack packet will change the state to $(N,2)$. Since router use a round-robin policy, each server has equal share of load. Therefore, the transition rate is λ/N.

- current state is $(N,2)$, departure of one packet will change the state to $(N,1)$. Since each server has equal processing rate of μ, therefore, the transition rate is μ/N.

101

- current state is $(N,2)$, one hardware failure will make the next state $(N-1,1)$. Since we assume the hardware failure always occurs on the servers with heaviest load (equals one in this case), the packets assigned to the failed server will be lost.

- current state is $(N-1,1)$, the repair of the failed server will change the state of $(N,1)$.

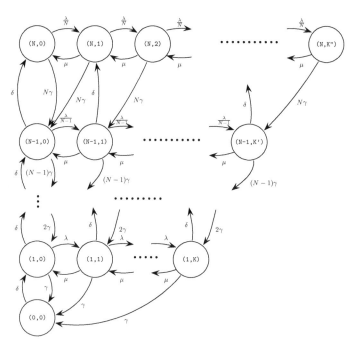

Figure 6.3: State digram of N location managers

We can determine each element of infinitesimal generator matrix Q of CTMC shown in Fig. 6.3 as follows:

$$q_{i,j} = \begin{cases} \lambda/N_w & j = i+1, L_i \leq N_w K \quad \text{(arrival)} \\ \mu & j = i-1, L_i \geq 1 \quad \text{(departure)} \\ \gamma N_w & j = i - \left\lceil \frac{i-1}{N_w} \right\rceil - \frac{K(N_w-1)}{2} \quad \text{(failure)} \\ \delta & j = i + N_w K + 1 \quad \text{(repair)} \\ 0 & \text{other } j \neq i \\ -\sum_{k=1}^{m} q_{i,k} & j = i, k \neq i \end{cases} \quad (6.1)$$

Where L_i is the total number of packets in system when current state is labelled as i, and m is the size of matrix, which is given by:

$$m = K \frac{N(N+1)}{2} + (N+1) \quad (6.2)$$

In the failure case in Eqn. 6.1, j is determined by:

$$j = \left(i - 1 - \sum_{x=0}^{N_w-1} \sum_{z=0}^{xK} 1 \right) - \left\lceil \frac{\left(i - 1 - \sum_{x=0}^{N_w-1} \sum_{z=0}^{xK} 1 \right)}{N_w} \right\rceil + \left(1 + \sum_{x=0}^{N_w-2} \sum_{z=0}^{xK} 1 \right)$$

$$= [i - (N_w - 1)K - 1] - \left\lceil \frac{\left(i - 1 - \sum_{x=0}^{N_w-1} \sum_{z=0}^{xK} 1 \right)}{N_w} \right\rceil$$

$$= i - \left\lceil \frac{i-1}{N_w} \right\rceil - \frac{K(N_w-1)}{2} \quad (6.3)$$

Once we have determined the infinitesimal generator matrix Q, we can compute the stationary distribution of the CTMC π by:

$$\pi Q = \mathbf{0} \quad (6.4)$$

When a packet arrives, if the system is in state (0,0) or a state where $(N_w, N_w K)$, the packet is dropped since no service is possible. Therefore, the blocking probability can be calculated by:

$$P_b = \pi B^T$$

where $B = [1, B_1, \cdots B_j \cdots B_N],$

and $B_j = [0, \cdots 0, 1]_{jK+1}, j = 1, \cdots, N$ (6.5)

The average number of packets in the whole system can be calculated by:

$$E[n] = \pi v^T$$

where $v = [v_0, v_1, \cdots v_j \cdots v_N],$

and $v_j = [0, 1, \cdots jK], j = 0, \cdots, N$ (6.6)

According to Little's law, the system response time can be determined by:

$$E[T] = \frac{E[n]}{\lambda_{accepted}} = \frac{E[n]}{\lambda(1 - P_b)} \qquad (6.7)$$

6.2.4 Analytical Model for MIP Survivability

In this section, the survivability of MIP is analyzed. We use the same assumptions and notations as used for SIGMA in Sec. 6.2.2. In addition to the notations in Sec. 6.2.2, let λ_d be the arrival payload data traffic rate at HA, then $\lambda = \lambda_u + \lambda_q + \lambda_a + \lambda_d$. Two modes of MIP will be considered here:

- *single server mode*: only one HA available for one network. Once failure happens, all service requests are blocked until the server repaired.

- *standby mode*: there are multiple HAs available, one of which is the primary HA. Once the primary HA fails, one of the backup HAs will be switched in within time T_{sw}. During T_{sw}, all service requests are blocked.

Both these two MIP modes can be modelled by a CMTC as shown in Fig. 6.4. At any time, there can only be at most one HA serving requests. Any hardware failure will move the state from $(1,L)$ $(L = 1, 2, \cdots, K)$ to $(0,0)$. In single server model, state $(0,0)$ models the time for server repair, whereas in standby mode, state $(0,0)$ models the time required for switching a standby server into primary one. Therefore, the value of δ in Fig. 6.4 can be determined as follows:

$$\delta = \begin{cases} \frac{1}{MTTR} & \text{(single server mode)} \\ \frac{1}{T_{sw}} & \text{(standby mode)} \end{cases} \tag{6.8}$$

From now on, we can use the same technique as used in Sec. 6.2.3 to compute the average system response time and service blocking probability by setting $N = 1$, and δ to the value given in Eqn. 6.8.

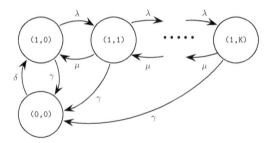

Figure 6.4: State digram of MIP HA

6.3 Numerical Results

In this section, we evaluate the survivability of SIGMA through the analytical model developed in 6.2. The survivability of SIGMA is also compared with that of

MIP. The survivability is measured by the combined performance index in terms of system response time and blocking probability.

6.3.1 SIGMA Survivability

First, we look at the impact of DDoS attack strength (λ_a) on the system response time. We set $N = 3$, $\lambda_u = 0.2$, $\lambda_q = 0.4$, $\mu = 2$, $1/\delta = 24$ hours, and $K = 10$ packets. We choose $MTTF$ ranging from 24 to 3000 hours since the continous running time of the most of the current DNS servers fall into this range. As shown in Fig. 6.5, when DDoS attack has a higher strength, the system response time increases dramatically to as high as four times of normal values. Also, when the hardware failure is more frequent (smaller MTTF values), the system response time also increases due to less working server available to process client requests.

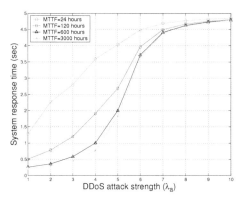

Figure 6.5: Impact of DDoS attack strength on system response time

Next, we look at the impact of DDoS attack strength on the system blocking probability. As shown in Fig. 6.6, when DDoS attack has a higher strength, the system blocking probability increases as well, due to less buffer space available to serve legitimate client requests. As expected, the smaller K is, the larger the

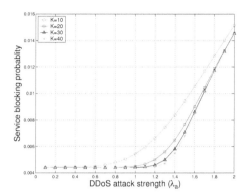

Figure 6.6: Impact of DDoS attack strength on service blocking probability

impact of DDoS attack on blocking probability. Therefore, increase the value of K can decrease the sensitivity of system blocking probability to DDoS attack. Fig. 6.7 shows the impact of MTTR on system response time. We can observe that the longer time repairing requires, the higher the average response time. This is because once a server fails, it needs longer time to repair it. Thus less working server is available to process client requests when MTTR is higher, which results in a higher response time.

Finally, Fig. 6.8 shows the impact of limiting availability on system response time. The limiting availability [49] is defined as $\alpha = \frac{MTTF}{MTTF+MTTR}$, which denotes the long range average percentage of available time. As expected, when α increase, the system response time decrease.

6.3.2 Survivability Comparison of SIGMA and MIP

Now, we compare the survivability of SIGMA against MIP. First, we look at the impact of DDoS attack strength (λ_a) on the system response time, with $\lambda_d = 0$, $T_{sw} = 10$ minutes and $MTTR = 24$ hours, as shown in Fig. 6.9. We can observe that the average response time in both modes of MIP is much higher than that

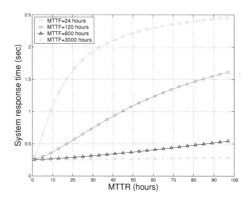

Figure 6.7: Impact of MTTR on system response time

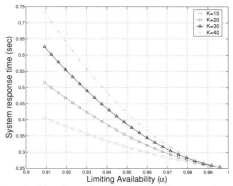

Figure 6.8: Impact of hardware limiting availability on system response time

of SIGMA, even with $\lambda_d = 0$. The value of MTTF does not have an impact on the response time for MIP. This is because we only consider the response time for non-blocked requests. Higher MTTF will results in system staying in available state more time, but more queueing delays will be incurred, these two effects are cancelled out, leaving no effect on the overall response time.

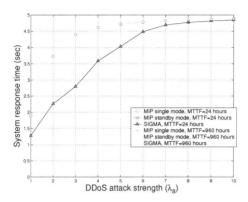

Figure 6.9: Impact of DDoS attack strength on system response time with zero λ_d and MTTR=24hr

Then we change $MTTR$ to be one hour and the resulting service response time is shown in Fig. 6.10, from which we can see that a lower $MTTR$ value will offset the impact of $MTTF$ value (the curve for $MTTF = 24$ and $MTTF = 960$ in Fig. 6.10 is much closer than in Fig. 6.9).

Next, we compare the impact of DDoS attack strength on the service blocking probability of SIGMA against MIP. As shown in Fig. 6.11, when DDoS attack has a higher strength, all schemes incur a higher service blocking probability. However, SIGMA has a lower blocking probability than both modes of MIP. For MIP standby mode, MTTF does not have obvious impact on service blocking probability. This is because that T_{sw} is 10 minutes, which is so small compared to MTTF. Once HA fails, it can be deemed as to be replaced by a new one immediately.

Again, the $MTTR$ is reduced to one hour to show the effect on service blocking probability in Fig. 6.12, from which we can also see that a lower $MTTR$ value will offset the impact of $MTTF$ value(the curve for $MTTF = 24$ and $MTTF = 960$ in Fig. 6.12 is much closer than in Fig. 6.11).

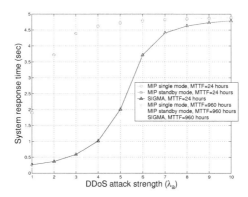

Figure 6.10: Impact of DDoS attack strength on system response time with $\lambda_d = 0$ and $MTTR$=1hr

Fig. 6.13 compare the impact of data traffic strength on the service blocking probability of SIGMA against MIP, with $\lambda_a = 1$. Since SIGMA decouples the location management from data forwarding, the data traffic strength does not have impact on the service blocking probability. For MIP, the data traffic will contend with location management traffic for the buffer slots, which will increase the blocking probability. This observation justifies our initial design of decoupling the location management from data forwarding function in SIGMA.

Fig. 6.14 compare the impact of hardware limiting availability on the response time of SIGMA against MIP. As in the case of MTTF in Fig. 6.9, the limiting availability does not affect the response time of MIP. Since MTTR is fixed, the limiting availability only depends on MTTF according to its definition. In comparison, higher α (which means server hardware is more reliable) will results a lower response time for SIGMA.

110

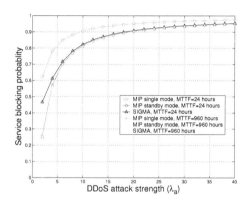

Figure 6.11: Impact of DDoS attack strength on service blocking probability with $\lambda_d = 0$ and $MTTR$=24hr

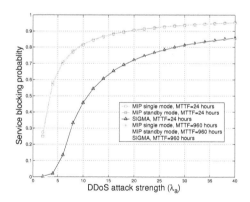

Figure 6.12: Impact of DDoS attack strength on service blocking probability with $\lambda_d = 0$ and $MTTR$=1hr

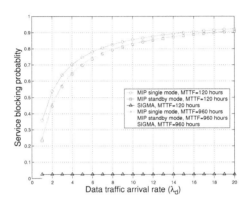

Figure 6.13: Impact of data traffic strength on blocking probability

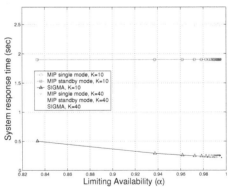

Figure 6.14: Impact of hardware limiting availability on system response time

6.4 Summary

In this chapter, we show that the location management scheme used in SIGMA can enhance the survivability of the mobile network. We developed an analytical model to evaluate the survivability of location management schemes. Through the model, the survivability of SIGMA as compared to that of Mobile IP. Numerical results have shown SIGMA has better survivability than Mobile IP in terms of system response time and service blocking probability, in practical environments under the risk of hardware failures and distributed DoS attacks.

Chapter 7

Hierarchical Location Management of SIGMA

The location management scheme presented earlier in Chapter 3 is not suitable for frequent mobile handovers due to user's high mobility. The reasons are as follows:

- There is a race condition between (Location Manager) LM database update caused by the change of MH's point of attachment and the arrival of association setup request from CN. The higher the Round Trip Time (RTT) between MH and LM is, the larger probability that CN get a stale information from the database at LM, which will result in MH being inaccessible from CN.

- Performing location update on LM whenever MH changes its location may be too costly and time-consuming for LM to process. Too many signaling messages exchanged in the network wastes network bandwidth and may result in unnecessary congestions.

- DNS servers commonly cache DNS replies to reduce the signaling load on network and response time to CN. Each DNS reply is associated with a Time-To-Live (TTL) field indicating the valid period of the cached DNS reply. During the TTL period, the DNS server with cache could answer additional requests for the MH's location from its local cache instead of querying LM again. Thus, even after MH has updated its location with LM, the CN's DNS server could still reply with the old location until the cached entry's TTL expire. This will also lead to MH being inaccessible from CN.

The *objective* of this chapter is to propose a hierarchical location management scheme for transport layer mobility solutions to reduce the possibility that MH is inaccessible from CNs and the processing load on LM. The contributions of this work can be outlined as follows:

- Propose and develop a hierarchical location management scheme for transport layer mobility protocols.

- Evaluate and compare the signaling cost of proposed the hierarchical management scheme with that of HMIPv6 [4] using analytical models.

The authors are not aware of any *previous studies for hierarchical location management for transport layer mobility solutions.* For instance, the signaling cost, which is a very important performance measures for a location management scheme, is not investigated by the authors of [6, 7, 63, 64]. The rest of this chapter is structured as follows: Sec. 7.1 describes the hierarchical location management scheme including its architecture, timeline, and state machine. The analytical model for HiSIGMA signaling cost is developed in Sec. 7.2. The results of signaling cost comparison of HiSIGMA and HMIPv6 is presented in Sec. 7.3.

7.1 Hierarchical Location Management of Transport Layer Mobility

In this section, we introduce hierarchical location management for transport layer mobility. Since we use SIGMA as the base architecture for introducing hierarchical location management, we call the proposed scheme as HiSIGMA. However, the principle of HiSIGMA also applies to other transport layer mobility solutions such as [6, 7].

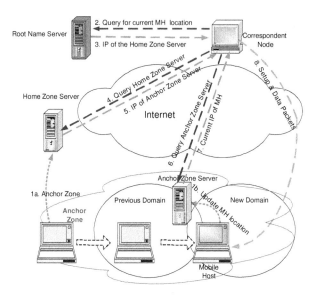

Figure 7.1: Hierarchical location management in HiSIGMA

7.1.1 Architecture of HiSIGMA

A new entity called Anchor Zone Server needs to be introduced in HiSIGMA as shown in Fig. 7.1. MH only needs to update the Home Zone Server when it enters a new Anchor Zone. Otherwise, MH need only to update the Anchor Zone Server with its current location. Whenever Home Zone Server receives a location query for MH, it will answer with the registered Anchor Zone Server's IP address. This approach will reduce the location update latency and signaling cost while improve the accuracy of the location management. The hierarchical location management can be done in the following sequence as shown in Fig. 7.1:

(1) a. When MH enters into a new DNS zone, MH updates the HZS with the IP address of new attached AZS. b. When MH moves between IP domains within the region managed by a specific AZS, MH only updates AZS.

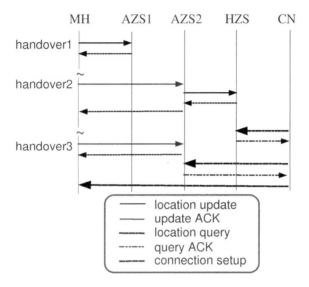

Figure 7.2: Time line of HiSIGMA

(2) When CN wants to setup a new association with MH, CN sends a query to the root name server with MH's domain name.

(3) Root name server replies to CN with the IP address of the HZS.

(4) CN query the HZS referred by the root name server.

(5) HZS replies with the IP address of current AZS where MH resides.

(6) CN query the AZS referred by the HZS.

(7) AZS replies with the current IP address(es) of MH.

(8) CN initiates the handshake sequence with MH's current IP address to setup the association.

The timeline for three handovers in HiSIGMA is shown in Fig. 7.2, where `handover1` and `handover3` are intra-AZS handovers within AZS1 and AZS2, respectively. And `handover2` is an inter-AZS handover which requires an update to HZS server. The signaling messages for CN querying MH's location and setting up a connection with MH are also shown in Fig. 7.2.

7.1.2 State machine at AZS

During the movement of MH, the IP address used by MH keeps changing. Furthermore, in schemes like SIGMA, the number of IP addresses that MH have also varies, sometimes one and sometimes two [65]. MH may also have its preference on which IP should be used at a particular time based on application characteristics (e.g. VoIP or data) and cost constraints (e.g. satellite links are generally more expensive than WLAN). To support this kind of desirable flexibility and optimize the performance of location management for transport layer mobility solutions that support IP diversity like SIGMA, a state machine is introduced at AZS. For the schemes in which mobile hosts do not support IP diversity, the hierarchical location management is still useful, but the lack of this state machine may result in non-optimal results.

It is necessary for AZS to have a clear idea on which IP address(es) should be used and which one has priority when multiple IP are available. In HiSIGMA, this goal is achieved by multicasting the IP reconfiguration information of MH to CN and AZS. When MH send IP reconfiguration signaling messages to CN, MH should also send a copy to AZS. These messages could include [65]:

- Add new IP into association between MH and CN (ADD_IP).

- Designate one of the available IP addresses as the primary destination address (SET_PRIMARY).

- Delete obsolete IP address (DELETE_IP).

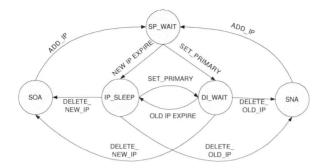

Figure 7.3: State machine at AZS

. These signaling messages are used to construct a state machine at AZS to better reflect the current location status of MH.

The state machine at AZS is shown in Fig. 7.3. The state machine works as follows:

- If MH has only one IP address assigned from the old domain or new domain, the AZS is in SOA (Single Old Address) or SNA (Single New Address) state, respectively.

- If current state is SOA or SNA, an ADD_IP message received from MH will trigger the machine to transfer into SP_WAIT state, which means that AZS is waiting for a SET_PRIMARY message.

- If current state is SP_WAIT or IP_SLEEP, a SET_PRIMARY message received from MH will trigger the machine to transfer into DI_WAIT state, which means that AZS is waiting for a DELETE_IP message.

- If current state is SP_WAIT, and the timer associated with the new IP just added into the association expires before a SET_PRIMARY message is received, the machine transfer into IP_SLEEP state, which means that the IP is marked as inactive and should not be advertised to CN.

119

- If current state is DI_WAIT or IP_SLEEP, and a DELETE_IP message is received from MH with the old IP address as the target IP being deleted, it will trigger the machine to transfer into SNA state. Similarly, if a DELETE_IP message is received with the new IP address as the target IP being deleted, it will trigger the machine to transfer into SOA state.

- If current state is DI_WAIT, and the timer associated with the old IP waiting to be deleted expires before a DELETE_IP message is received, the machine transfer into IP_SLEEP state, which means that the old IP is marked as inactive and should not be advertised to CN.

7.1.3 Location Query Replies Sent to CN by AZS

One of the most important objectives of location management is to accurately pointer CN to the current location of MH. We utilize the sate machine at AZS to improve this accuracy. The reply sent by AZS to CN depends on the current state of AZS as described below.

- *SOA or SNA*: Only one IP available at MH, just send MH's IP to CN.

- *SP_WAIT*: Send both MH's new and old IP to CN, old IP has higher priority.

- *DI_WAIT*: Send both MH's new and old IP to CN, new IP has higher priority.

- *IP_SLEEP*: Only one IP active at MH, send current MH's active IP to CN.

When CN receives a location reply with multiple entries of MH's IP address, it will first try the first entry. If the association setup using first entry fails, CN will automatically try the second entry.

7.2 Analytical Model of HiSIGMA Signaling Cost

In this section, we analyze the signaling cost of HiSIGMA using analytical model. The assumptions, notations, mobility model, and arrival traffic model is the same as used in Chapter 5. First, the network structure being considered in the model are presented in Sec. 7.2.1. Secs. 7.2.2, 7.2.3, and 7.2.4 develop the cost for location update, binding update and packet delivery, respectively. Sec. 7.2.5 gives the total signaling cost of HiSIGMA.

7.2.1   Network Structure

Fig. 7.4 shows a two dimensional subnet arrangement for modeling MH movement, where $AR_{1,1}, \cdots AR_{m,n}$ represent access routers. There are k AZSs, each of which covers R subnets. There are also one HZS (same as HA in the case of HMIPv6) and a number of CNs connected to the Internet. The MHs are roaming in the subnets covered by $AR_{1,1}, \cdots AR_{m,n}$, and each MH communicates with one or more of the CNs. Between a pair of MH and CN, intermittent file transfers occur caused by mobile users requesting information from CNs using protocols like HTTP. We call each active transfer period during the whole MH-CN interactivity as a session.

7.2.2   Location Update Cost

In HiSIGMA, an MH does not need to register with the HZS until the MH moves out of the region covered by an AZS, instead it only registers with the AZS. Therefore, every subnet crossing within a AZS (happens every T_r seconds) will trigger a registration to the AZS, which incurs a transmission cost to AZS (LU_{ma}) and processing cost at AZS (γ_a) of the location update message. Also, MH needs to update its current AZS with its dynamic address configuration messages to maintain the state machine at the AZS. Therefore, $C_{ma} = LU_{ma} + \gamma_a + AU_{ma}$.

For every region crossing between AZSs (happens every $M \times T_r$ seconds), MH needs to register with HZS, which incurs a transmission cost to HZS (LU_{ml}), processing cost at HZS (γ_l), and processing cost at AZS ($2\gamma_a$, since AZS needs to process both registration request and reply messages). Therefore, $C_{ml} = LU_{ml} + \gamma_l + 2\gamma_a$.

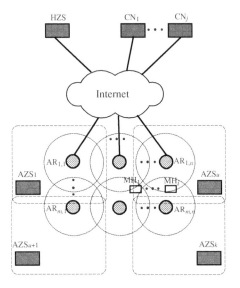

Figure 7.4: Network structure considered.

Since there is only one location update per subnet crossing, no matter how many CNs an MH is communicating with, the number of CNs does not have any impact on the location update cost. Therefore, the average location update cost per second in the whole system can be estimated as the number of MHs multiplied by the location update cost for each MH:

$$\Psi_{LU}^{T} = N_{mh}\frac{M^{T}C_{ma} + C_{ml}}{M^{T}T_{r}} \qquad (7.1)$$

M^{T} can be calculated from the total number of subnets ($m \times n$) and the number of subnets beneath a AZS (R^{T}): [50]:

$$M^{T} = 1 + \frac{mn - 1}{mn - R^{T}} \qquad (7.2)$$

Due to frame retransmissions and medium access contentions at the data link layer of wireless links, transmission cost of a wireless hop is higher than that of a wired hop; we denote this effect by a proportionality constant, ρ. Let the per-hop location update and dynamic address reconfiguration transmission cost be δ_U and δ_A, respectively. For a round trip, LU_{ma}, AU_{ma} and LU_{ml} can be calculated as:

$$LU_{ma} = 2(l_{ma} - 1 + \rho)\delta_U \qquad (7.3)$$

$$AU_{ma} = 2(l_{ma} - 1 + \rho)\delta_A \qquad (7.4)$$

$$LU_{ml} = 2(l_{ma} + l_{al} - 1 + \rho)\delta_U \qquad (7.5)$$

Where $(l_{ml} - 1)$ represents the number of wired hops. Therefore,

$$\Psi_{LU}^T = N_{mh} \left[\frac{2(l_{ma} - 1 + \rho)(\delta_U + \delta_A) + \gamma_a}{T_r} + \right.$$
$$\left. \frac{2(l_{ma} + l_{al} - 1 + \rho)\delta_U + \gamma_l + 2\gamma_a}{T_r} \times \frac{mn - R^T}{2mn - R^T - 1} \right] \qquad (7.6)$$

7.2.3 Binding Update Cost

In the analysis of binding update cost, processing costs at the endpoints (MH and CN) are not counted into the total signaling cost, since these costs stand for the load that can be scattered into user terminals and hence do not contribute to the network load. Because we are more concerned about the load on the network elements, this assumption enables us to concentrate on the impact of the handover protocol on network performance. This same assumption was also made by other previous works [50, 51, 55].

Similar to the analysis in Sec. 7.2.2, every subnet crossing will trigger a binding update to CN, which incurs a transmission cost (BU_{mc}) due to the binding update message. For each CN communicating with an MH, the MH need to send a binding update after each handover. Therefore, the average binding update cost can be estimated as:

$$\Psi_{BU}^{T} = N_{mh}N_{cn}\frac{BU_{mc}}{T_r} \qquad (7.7)$$

Let the per-hop binding update transmission cost be δ_B. The BU_{mc} can be calculated as:

$$BU_{mc} = 2(l_{mc} - 1 + \rho)\delta_B \qquad (7.8)$$

Therefore, the binding update cost per second in the whole system can be calculated by multiplying the number of MHs, the average number of communicating CNs, and the average cost per binding update:

$$\Psi_{BU}^{T} = N_{mh}N_{cn}\frac{2(l_{mc} - 1 + \rho)\delta_B}{T_r} \qquad (7.9)$$

7.2.4 Packet Delivery Cost

Unlike the analysis of packet delivery cost in [50], we do not consider the data packet transmission cost, IP routing table searching cost, and bandwidth allocation cost since these costs are incurred by standard IP switching, which are not particularly related to mobility protocols. Instead, we only consider the location database lookup cost at HZS and AZS. Moreover we take into account the processing cost caused by packet tunnelling to better reflect the impact of mobility protocol on overall network load.

For HiSIGMA, a location database lookup at HZS is required when an association is being setup between CN and MH. If each session duration time is independent from each other, the association setup event happens every S/λ_{sa} seconds. If we assume the database lookup cost has a linear relationship with N_{mh}, and φ_l and ψ be the per location database lookup cost and the linear coefficient at HZS, then the per-second per-association lookup cost v_l can be calculated as:

$$v_l = \frac{\varphi_l\lambda_{sa}}{S} = \frac{\psi N_{mh}\lambda_{sa}}{S} \qquad (7.10)$$

Let φ_a and ψ be the per location database lookup cost and the linear coefficient at AZS, then the per-second per-association lookup cost υ_a can be calculated as:

$$\upsilon_a = \frac{\varphi_a \lambda_{sa}}{S} = \frac{\psi N_{mh} R^T \lambda_{sa}}{mnS} \tag{7.11}$$

Since HiSIGMA is free of packet encapsulation or decapsulation, there is no processing cost incurred at intermediate routers. So the packet delivery cost from CN to MH can be calculated by only counting the location database lookup cost. This cost can be expressed as:

$$\begin{aligned}
\Psi_{PD}^T &= N_{mh} N_{cn}(\upsilon_l + \upsilon_a) \\
&= N_{mh}^2 N_{cn} \tfrac{\psi \lambda_{sa}}{S}(1 + \tfrac{R^T}{mn})
\end{aligned} \tag{7.12}$$

7.2.5   Total Signaling Cost of HiSIGMA

Based on above analysis on the location update cost, binding update cost, and packet delivery cost shown in Eqns. (7.6), (7.9), and (7.12), we can get the total signaling cost of HiSIGMA as:

$$\Psi_{TOT}^T = \Psi_{LU}^T + \Psi_{BU}^T + \Psi_{PD}^T \tag{7.13}$$

7.3   Results and Signalling Cost Comparison of HiSIGMA and HMIPv6

In this section, we present results showing the effect of various input parameters on HiSIGMA's total signaling cost. In all the numerical examples, using the following parameter values, which are obtained from previous work [50] and our calculation based on user traffic and mobility models [52,54]: $\gamma_l = \gamma_h = 30$, $\gamma_a = \gamma_m = 20$, $\psi = 0.3$, $F = 10$Kbytes, $PMTU = 576$bytes, $S = 10$, $\rho = 10$, $l_{al} = l_{mh} = 25$, $l_{ma} = l_{mm} = 10$, $l_{mc} = 35$, $m = 10$, $n = 8$, $R^T = R^H = 10$, $\tau = 0.5$, and $\lambda_{sa} = 0.01$.

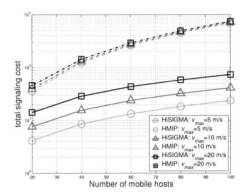

Figure 7.5: Impact of number of MHs on total signaling cost of HiSIGMA and HMIPv6 under different moving speeds.

7.3.1 Impact of Number of MHs for Different Moving Speeds

The impact of number of MHs on total signaling cost of HiSIGMA and HMIPv6 for different moving speeds is shown in Fig. 7.5. Here, the values used for other parameters are: $N_{cn} = 1$ and $\delta_U = \delta_B = \delta_A = 0.2$. From the figure, we can see that under different moving speeds, the signaling cost of both HiSIGMA and HMIPv6 increases with the increase of the number of MHs. When the moving speed is higher, the subnet residence time T_r decreases, resulting in a increase of the location update and binding update costs per second (see Eqns. (7.6) and (7.9) and (5.29)). We can also observe that the total signaling cost of HiSIGMA is less than HMIPv6 in this scenario; this is because when δ_U and δ_B are small, the location update and binding update costs are not high, and the high packet delivery cost will make the signaling cost of HMIPv6 much higher than that of HiSIGMA.

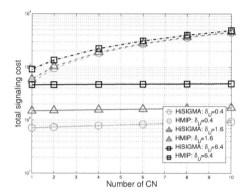

Figure 7.6: Impact of number of CNs and per-hop binding update transmission cost

7.3.2 Impact of Average Number of Communicating CN and Location Update Transmission Cost

Next, we set MH maximum moving speed $v_{max} = 20m/s$, and number of MHs $N_{mh} = 80$. The impact of the number of average CNs with which an MH communicates with for different per-hop transmission cost for location update cost (δ_U) is shown in Fig. 7.6. It can be observed from this figure that when the average number of communicating CNs increases, the total signaling cost increases (see Eqns.(7.6), (7.9) (7.12),(5.29) and (5.32)).

Also, when δ_U increases, the location update cost per second will increase as indicated by Eqn. (7.5), (5.26) and (5.27), which will result in the increase of the total signaling cost of both HiSIGMA and HMIPv6. However, we can see that the impact of δ_U is much smaller in HMIPv6; this is because HMIPv6's signaling cost is less sensitive to location update cost due to its hierarchical structure. In this scenario, signaling cost of HMIPv6 is higher than that of HiSIGMA when $\delta_U = 0.4$ or 1.6. However, when $\delta_U = 6.4$, HiSIGMA requires a higher signaling cost

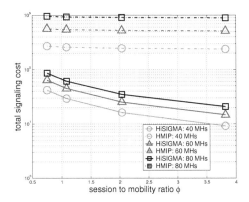

Figure 7.7: Impact of SMR on total signaling cost for different N_{mh}

due to frequent location update for each subnet crossing (compared to HMIPv6's hierarchical mobility management policy).

7.3.3 Session to Mobility Ratio

Session to Mobility Ratio (SMR) is a mobile packet network's counterpart of Call to Mobility Ratio (CMR) in PCS networks. We vary T_r from 75 to 375 seconds with λ_{sa} fixed to 0.01, which yields a SMR of 0.75 to 3.75. The impact of SMR on total signaling cost for different N_{mh} is shown in Fig. 7.7. We can observe that a higher SMR results in lower signaling cost in both HiSIGMA and HMIPv6. This is mainly because high SMR means lower mobility, and thus lower signaling cost due to less location update and binding update.

7.3.4 Relative Signaling Cost of HiSIGMA to HMIPv6

Fig. 7.8 shows the impact of (location update transmission cost) / (packet tunnelling cost) ratio (δ_U/τ) on the relative signaling cost between HiSIGMA and

HMIPv6. A higher δ_U/τ ratio means that the location update requires more cost while packet encapsulation/decapsulation costs less. This ratio depends on the implementation of the intermediate routers. We can see that the signaling cost of HiSIGMA is less than that of HMIPv6 in the possible range of δ_U/τ since the relative cost between HiSIGMA and HMIPv6 is always less than one.

7.4 Summary

In this chapter, we presented the hierarchical location management scheme for SIGMA– HiSIGMA. We developed an analytical model to evaluate HiSIGMA using signaling cost as the performance measure, followed by a comparison of the signalling cost of HiSIGMA and Hierarchical Mobile IPv6. Numerical results show that, by introducing the concept of Anchor Zone Server into location management of mobile hosts, the signaling cost of HiSIGMA can be greatly reduced and is lower than that of HMIPv6.

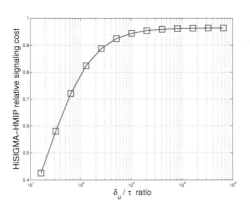

Figure 7.8: Impact of δ_U/τ ratio on HiSIGMA to HMIPv6 relative signaling cost

Chapter 8

SIGMA-SN: Applying SIGMA to Satellite Handovers in Space Networks

Satellite networks have a large coverage area, and currently provide television, radio, telephony, and navigation services. Satellites are expected to play a significant role in the future global Internet to provide broadband data services. Satellites communicate among themselves and with ground stations on the earth to enable space communications.

Depending on the altitude, satellites can be classified into three types: Low Earth Orbit (LEO), Medium Earth Orbit (MEO) and Geosynchronous Earth Orbit (GEO) [66]. GEO satellites are stationary with respect to earth, but LEO and MEO satellites move around the earth, and are handed over between ground stations as they pass over different areas of the earth. This is analogous to mobile computers being handed over between access points as the users move in a terrestrial network. Currently, GEO and LEO satellites are mostly used for the Internet applications.

Traditionally, GEO satellites have been used to provide a *bent-pipe* transmission channel, where all packets received on an up-link are transparently piped to the corresponding down-link, i.e. a GEO satellite is merely a physical layer repeater in space, which is invisible to the routing protocols. To increase system capacity and reduce end-to-end delay, newer satellites are increasingly adopting a regenerative paradigm where the satellites have on-board switching and routing units [67]. This is also consistent with the current efforts of the National Aeronautics and Space Administration [68] and the European Space Agency [69].

The long propagation delay of GEO satellites make them less desirable for real-time applications, such as voice communications. The concept of LEO satellite constellation was introduced in 1990s to provide satellite services at a lower obit by utilizing a large number of satellites than a GEO constellation. The advantages of LEO over GEO include lower link propagation delay, reduced free space attenuation, lower power consumption for user terminals, and higher spectrum efficiency due to frequency reuse [66]. However, these advantages come at the cost of a large number of satellites required to be launched and maintained (even though a LEO satellite is less expensive than a GEO one). Additionally, mobility management issues, arising due to the non-stationary nature of LEO satellites with respect to the Earth, have to be to considered.

The National Aeronautics and Space Administration (NASA) has been studying the use of Internet protocols in spacecrafts for space communications [70]. For example, the Global Precipitation Measurement (GPM) project is studying the possible use of Internet technologies and protocols to support all aspects of data communication with spacecraft [71]. The Operating Missions as Nodes on the Internet (OMNI) [72, 73] project at GSFC is not only involved in prototyping, but is also testing and evaluating various IP-based approaches and solutions for space communications. Other efforts in using Internet protocols for space communications have also been reported in the literature [74].

Some of the NASA-led projects on IP in space involve handoffs in space networks. Such projects include OMNI [72, 75], Communication and Navigation Demonstration on Shuttle (CANDOS) mission [76], and the GPM project [77]. NASA has also been working with Cisco on developing a Mobile router [78]. It is also anticipated that MIP will play a major role in various space related NASA projects such as Advanced Aeronautics Transportation Technology (AATT), Weather

132

Information Communication (WINCOMM) and Small Aircraft Transportation Systems (SATS) [78].

In this chapter, we will investigate the use of SIGMA in space networks to support IP mobility. First, effects of satellite link characteristics on transport protocols are discussed in Sec. 8.1. Two types of satellite constellations are introduced in Sec. 8.2. The scenarios of network layer handovers in satellite environment is identified in Sec. 8.3. Then, we introduce SIGMA-SN — the mapping of SIGMA in space network.

8.1 Effects of Satellite Link Characteristics on Transport Protocols

A number of satellite link characteristics, which are different from terrestrial links, may limit the performance of transport protocols over satellite networks [79, 80]. Because SCTP and TCP use similar congestion control, retransmission, and round trip time estimation algorithms, the characteristics have many similar effects on the two protocols. The following are the satellite link characteristics which are of interest in this paper.

- *Long propagation delay*: The propagation delay between an earth station and a GEO satellite is around 120-140ms (milliseconds), which means that it takes the sender a long time to probe the network capacity and detect possible loss of segments, resulting in expensive satellite bandwidth being wasted.

- *Large delay-bandwidth product*:The GEO satellite link is a typical case of the Long Fat Pipe (LFP), which features a large delay bandwidth product. For example, the DS1-speed GEO channel has a 96500-byte size pipe. The fundamental performance problems with the current TCP over LFN links were discussed in [81].

- *Errors due to propagation corruption and handovers*: The frequent fading of satellite links results in a low signal-to-noise ratio (SNR) and consequently a high Bit Error Rate (BER) during free space propagation. The GSL handovers in LEO constellations will also contribute to the burst errors observed by the endpoints. These errors will cause TCP and SCTP senders to activate congestion control mechanisms, and reduce their transmission rates unnecessarily.

- *Variable Round Trip Time and Link handovers*: The ground stations in LEO satellite system generally experience a handover interval of only a few minutes between two satellites. Propagation delay between ground and LEO varies rapidly as a satellite approaches and leaves a ground station. During the handover, packets can experience a much higher RTT than during normal periods. Transport layer protocols, like TCP and SCTP, depend on accurate RTT estimation to perform congestion control; too frequent RTT change may cause problems for TCP RTO calculation algorithms.

Although TCP is the dominant transport protocol in the IP protocol suite, it was not initially designed for long bandwidth delay product networks, such as satellite networks which are characterized by long propagation delays and corruption losses due to wireless links. Consequently, enhancements to improve the performance of TCP over satellite networks have been proposed [79, 80, 82]. In one of our previous papers [14], we evaluated and recommended SCTP features that can be exploited to increase SCTP's performance over satellite networks.

8.2   Illustration of Satellite Constellations

We consider two types of satellite constellations: a GEO constellation proposed by the Clarke model [83], and a LEO constellation called Iridium [84, 85]. The GEO constellation resides at an altitude of 35786 km, and each satellite has onboard processing capability to route the packets. We choose Iridium as the LEO

Figure 8.1: Mixed constellation of Iridium and GEO.

constellation in this paper because it is the first operational LEO system that provides truly global coverage. Fig. 8.1 shows the satellites and their orbits in both a GEO and Iridium LEO constellation.

The Iridium constellation consists of 66 satellites, grouped into 6 *planes* with each plane have 11 satellites. Each satellite has four 25Mbps inter-satellite links (ISL), which operate in the frequency range of 22.55 to 23.55 GHz. Two of the ISLs (called *intraplane* ISL) connect a satellite to its adjacent satellites in the same plane, and the other two ISLs (*interplane* ISL) connect it to the satellites in the neighboring co-rotating planes. The inter-plane ISLs is temporarily deactivated near the poles because of antenna limitations in tracking these ISLs in polar areas [66]. Each earth endpoint can be connected to a GEO and/or a LEO satellite through ground-to-satellite link (GSL). In the case of connection to LEO, the GSL links experience periodical handovers to accommodate the relative movement of

135

Table 8.1: Orbit and link characteristics of GEO constellation.

Number of satellites	3
Altitude	35786.1 km
Longitude	-90°, 30°, 150°
Period time	24 hours
GSL link bandwidth	2Mbps
Path BER	10^{-4} to 10^{-9}

Table 8.2: Orbit and link characteristics of Iridium constellation.

Number of planes	6
Number of satellites/plane	11
Altitude	780.0 km
Geometry	polar orbits at 86.4^o incl.
Period time	100.4 minutes
Interplane separation	31.6^o
Minimum elevation angle	8.2^o
ISLs per satellite	4
GSL link bandwidth	1.5Mbps
ISL link bandwidth	25Mbps
Path BER	10^{-4} to 10^{-9}

the LEO satellites and the Earth. The orbit and link characteristics of the GEO and LEO satellite constellations are shown in Tables 8.1 and 8.2, respectively.

Fig. 8.2 shows a complete snapshot of the ISLs and GSLs in an Iridium-GEO constellation through which 30 SCTP associations between 30 pairs of end points are setup. The six planes of the Iridium constellation are shown by the nearly-vertical lines (since Iridium's inclination is 86.4^o). Each LEO satellite has four ISLs: two intraplane and two interplane. Since the satellites in the two planes near 0° longitude are counter rotating, there is no interplane ISL between the two planes near 0° longitude. Three GEO satellites reside at longitudes of -90°, 30°, and 150°. Since each of the GEO satellites has, on the average, 20 GSLs setup to support 30 SCTP associations requiring 60 GSLs, the GSLs connected to the three GEO satellites appear to be denser as compared to the LEO GSLs.

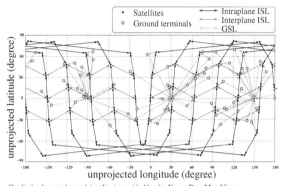

Credit: background map 1 (outline) provided by the Xerox Parc Map Viewer
background maps 2 and 3 (grayscale photo montages) provided as samples by Living earth
background map 4 (land masses) courtesy of the footprint generator from L. Wood

Figure 8.2: Unprojected map of mixed Iridium/GEO constellation

8.3 Handovers in a Satellite Environment

LEO satellites have some important advantages over GEO satellites for implementing IP in space. These include lower propagation delay, lower power requirements both on satellite and user terminal, more efficient spectrum allocation due
to frequency reuse between satellites and spotbeams. However, due to the non-
geostationary nature and fast movement of LEO satellites, the mobility management in LEO is much more challenging than in GEO or MEO.

If one of the communicating endpoint (either satellite or user terminal) changes
its IP address due to the movement of satellite or mobile user, a network layer
handoff is required to migrate the connection of higher level protocol (e.g. TCP,
UDP, or SCTP) to the new IP address. We describe below two scenarios requiring
network layer handoff in a satellite environment.

137

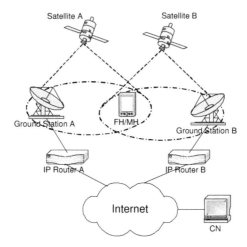

Figure 8.3: User handoff between satellites.

8.3.1 Satellite as a Router

When a satellite does not have any on-board equipment which generates or consumes data, but is only equipped with on-board IP routing devices, the satellite acts as a router in the Internet. Hosts are handed over from one satellite to another as the hosts come under the footprint of different satellites due to the rotation of the LEO satellites around the Earth. Referring to Fig. 8.3, the Fixed Host/Mobile Host (FH/MH) needs to maintain a continuous transport layer connection with the correspondent node (CN) while their attachment points change from Satellite A to Satellite B. Different satellites, or even different spotbeams within a satellite, can be assigned with different IP subnet addresses. In such a case, IP address change occurs during an inter-satellite handoff, thus requiring a network layer handoff. For highly dense service areas, a spot-beam handoff may also require a network layer handoff. For inter-satellite handoff, the two satellites could be both LEO satellites, or one LEO satellite and another GEO or MEO satellite. Previous research [86,87]

138

have used Mobile IPv6 to support mobility management in LEO systems, where the FH/MH and Location Manager are mapped to Mobile IP's Mobile Node and Home Agent, respectively.

8.3.2 Satellite as a Mobile Host

When a satellite has on-board equipment (such as earth and space observing equipment) which generates data for transmission to workstations on Earth, or the satellite receives control signals from the control center, the satellite acts as the endpoint of the communication, as shown in Fig. 8.4. Although the satellite's footprint moves from ground station A to B, the satellite should maintain continuous transport layer connection with its corespondent node (CN). A network layer handoff has to be performed if the IP address of the satellite needs to be changed due to the handover between ground stations.

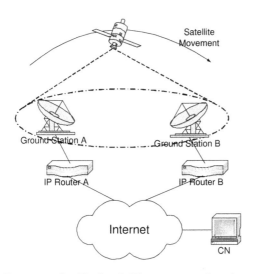

Figure 8.4: Satellite handoff between ground stations.

8.4 SIGMA-SN: SIGMA in Space Networks

Having described our proposed SIGMA architecture and handovers in space networks in Chapter 3 and Sec. 8.3, respectively, we describe below the mapping of SIGMA into a space handoff scenario, using satellites as examples of spacecrafts. We call this application and mapping of SIGMA to a space environment as SIGMA-SN.

8.4.1 Satellite as a Router

Research results desribed in [88] showed that the mean number of available satellites for a given FH/MH is at least two for latitudes less than 60 degrees. This means the FH/MH is within the footprint of two satellites most of the time, which makes SIGMA-SN very attractive for handoff management with a view to reducing packet loss and handoff latency. The procedure of applying SIGMA in this handoff scenario is straightforward; we just need to map the FH/MH and satellites in Fig. 8.3 to the MH and access routers, respectively, in the SIGMA scheme (see Fig. 3.5) as given below:

- *Obtain new IP*: When FH/MH receives advertisement from Satellite B, it obtains a new IP address using either DHCP, DHCPv6, or IPv6 Stateless Address Autoconfiguration.

- *Add new IP address to the association*: FH/MH binds the new IP address into the association (in addition to the IP address from Satellite A domain). FH/MH also notifies CN about the availability of the new IP address by sending an ASCONF chunk [36] to the CN with the parameter type set as "Add IP Address".

- *Redirect data packets to new IP address*: CN can redirect data traffic to the new IP address from Satellite B to increase the possibility of data being delivered successfully to the FH/MH. This task can be accomplished by sending an ASCONF chunk with the Set-Primary-Address parameter to CN, which results in CN setting its primary destination address to FH/MH as the new IP address.

- *Updating the Location manager*: SIGMA-SN supports location management by employing a location manager that maintains a database which records the correspondence between FH/MH's identity (such as domain name) and its current primary IP address. Once the Set-Primary-Address action is completed successfully, FH/MH updates the location manager's relevant entry with the new IP address. The purpose of this procedure is to ensure that after FH/MH moves from the footprint of Satellite A to that of Satellite B, further association setup requests can be routed to FH/MH's new IP address.

- *Delete or deactivate obsolete IP address*: When FH/MH moves out of the coverage of satellite A, FH/MH notifies CN that its IP address in Satellite A domain is no longer available for data transmission by sending an ASCONF chunk to CN with parameter type "Delete IP Address".

Due to the fixed movement track of the satellites in a space environment, FH/MH can predict the movement of Satellites A and B quite accurately. This a-priori information will be used to decide on the times to perform the set primary to the new IP address and delete the old IP address. This is much easier than in cellular networks, where the user mobility is hard to predict precisely.

8.4.2   Satellite as a Mobile Host

In this case, the satellite and IP Router A/B (see Fig. 8.4) will be mapped to the MH and access routers, respectively, of SIGMA. In order to apply SIGMA-SN, there is no special requirement on the Ground Stations A/B and IP routers A/B in Fig 8.4, which will ease the deployment of SIGMA-SN by not requiring any change to the current Internet infrastructure. Here, the procedure of applying SIGMA-SN is similar to the previous case (where the satellite acts as a router) if we replace the FH/MH by the satellite, in addition to replacing Satellites A/B by IP routers A/B.

Since a satellite can predict its own movement track, it can contact Ground Station B while it is still connected to Ground Station A. There may be multiple new Ground Stations available to choose from due to the large footprint of satellites. The strategy for choosing a Ground Station can be influenced by several factors, such as highest signal strength, lowest traffic load, and longest remaining visibility period.

8.5 Vertical Handoff between Heterogeneous Technologie

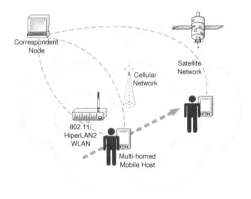

Figure 8.5: Vertical handoff using SIGMA-SN.

Different types of wireless access network technologies can be integrated to give mobile users a transparent view of the Internet. Handoff will no longer be limited to between two subnets in Wirless LAN (WLAN), or between two cells in a cellular network (horizontal handoff). In the future, mobile users will expect seamless handoff between heterogeneous access networks (vertical handoff), such as WLANs and cellular networks.

MIP operates in Layer 3 and is independent of the underlying access network technology. Although it can be used for handoffs in a heterogeneous environment, there are a number of disadvantages in using MIP for vertical handoffs [89]. The disadvantages include complexity in routing, high signaling overhead, significant delay especially when CN is located in foreign network, difficulty in integrating QoS protocols such as RSVP with triangular routing and tunnelling.

SIGMA-SN is well suited to meeting the requirements of vertical handoff. Fig. 8.5 illustrates the use of SIGMA-SN to perform vertical handoffs from WLAN to a cellular network, and then to a satellite network. A multi-homed mobile host in SIGMA-SN is equipped with multiple interface cards that can bind IP addresses obtained from different kinds of wireless network access technologies.

8.6   Summary

We have described the various types of handovers that can occur in space networks and how SIGMA-SN can be used to manage those handovers in the space environment. We have shown that various components of SIGMA-SN can be directly mapped to the architectural elements of the space handover scenario. We conclude that SIGMA-SNis suitable for managing handovers in space networks.

Chapter 9

Interoperability of SIGMA with Internet Security Infrastructure

MIP is known to have conflict with network security solutions [19]. Base MIP has difficulty in the presence of a foreign network which implements ingress filtering, unless reverse tunnelling, where the HA's IP address is used as the exit point of the tunnel, is used to send data from the MH.

The *objective* of this chapter is to show that the security of SIGMA can be enhanced by IP Security (IPSec) without conflicting with the existing Internet security features such as Ingress filtering.

9.1 Interoperability between MIP and Ingress Filtering

Ingress filtering is widely used in the Internet to prevent IP spoofing and Denial of Service (DoS) attacks. Ingress filtering is performed by border routers to enforce topologically correct source IP address. Topological correctness requires MH to use COA as the source IP address, since the COA is topologically consistent with the current network of the MH. On the other hand, TCP keeps track of its internal session states between communicating endpoints by using the IP address of the two endpoints and port numbers [18]. Therefore, applications built over TCP require the MH to always use its home address as its source address. The solution to this contradiction caused by combined requirements of user mobility, network security and transport protocols is *reverse tunnelling*, which works but lacks in terms of performance as illustrated below.

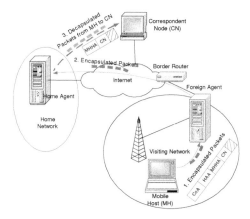

Figure 9.1: Interoperability between Mobile IP and Ingress Filtering.

Reverse tunnelling in MIP is shown in Fig. 9.1 which consists of the following components [1]:

(1) *Encapsulation*: A data packet sent from the MH to the CN has two IP headers: the inner header has source IP address set to MH's home address (MHHA) and destination IP address set to CN's IP address; the outer header has its source IP address set to MH's CoA and destination IP address set to HA's IP address (HAA). Since the MH's CoA is topologically correct with the foreign network address, ingress filtering at foreign network's border routers allows these packets to pass through.

(2) *Decapsulation*: The packets from the MH are routed towards the MH's HA because of the outer IP destination address. The HA decapsulates the packets, resulting in data packets with only one IP header (same as the previous inner header), which are then forwarded to their actual destination, i.e. the CN.

(3) *Data Delivery*: When data packets arrive at the CN with the source and destination addresses being that of MH's home address and CN's address, respectively, they are identified by its TCP connection and delivered to the upper layer application.

Reverse tunnelling makes it possible for MIP to interoperate with Ingress filtering. However, the encapsulation and decapsulation of packets increase the end-to-end delay experienced by data packets, and also increase the load on the HA, which may become a performance bottleneck as the number of MHs increases.

9.2 Interoperability between SIGMA and Ingress Filtering

In SIGMA, the transport protocol uses IP diversity to handle multiple IP addresses bound to one association. The CN can thus receive IP packets from multiple source IP addresses belonging to an association, identify the association, and deliver the packets to the corresponding upper layer application. This improved capability of endpoint transport protocol permits smooth interoperability between SIGMA and Ingress Filtering.

As shown in Fig. 9.2, MH can use the CoA that belongs to the subnet which is responsible for sending data for the MH. In the new network, after the new CoA (NCoA) has been bound into the current association through ADDIP chunks (discussed in Sec. 3.3), the MH uses the NCoA to communicate directly with the

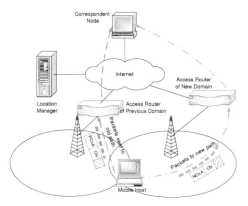

Figure 9.2: Interoperability between SIGMA and Ingress Filtering.

CN. Since the NCoA is topologically correct with the subnet address, the border router of the foreign network allows packets with source IP set to the new CoA to pass. Thus, SIGMA does not require encapsulation and decapsulation as done in MIP. The transport protocol stack at the CN takes care of delivering packets coming from both previous CoA (PCoA) and NCoA to the upper layer application. SIGMA, therefore, interoperates well with ingress filtering without the need for reverse tunnelling.

9.3 Enhancing the Security of SIGMA by IPSec

IPSec has been designed to provide an interoperable security architecture for IPv4 and IPv6. It is based on cryptography at the network layer, and provides security services at the IP layer by allowing endpoints to select the required security protocols, determine the algorithms to use, and exchange cryptographic keys required to provide the requested services. The IPSec protocol suite consists of two security protocols, namely Authentication Header (AH) and Encapsulating Security

147

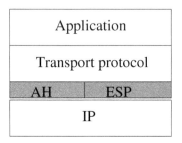

Figure 9.3: Use of IPSec with SCTP.

Payload (ESP). ESP provides data integrity, authentication, and secrecy services, while the AH is less complicated and thus only provides the first two services. The protocol stack, when IPSec is used with a transport protocol (SCTP in our case), is shown in Fig. 9.3.

SIGMA is based on dynamic address reconfiguration, which makes the association vulnerable to be hi-jacked, also called *traffic redirection attack*. An attacker claims that its IP address should be added into an established association between MH and CN, and further packets sent from CN should be directed to this IP address. Another kind of security risk is introduced by dynamic DNS update. An attacker can send a bogus location update to the location manager, resulting in all future association setup messages being sent to illegal IP addresses. The extra security risk introduced by SIGMA gives rise to the authentication problem: CN and LM need to determine whether the MH initiated the handover process. Since both AH and ESP support authentication, in general, we can choose either of them for securing SIGMA. ESP has to be used if data confidentiality is required. Assume that we are only concerned with authentication of MH by CN and LM to prevent redirection attack and association hi-jacking. In this case, AH can be used as shown in Fig. 9.4. All address reconfiguration messages and location updates sent to CN and LM should be protected by IPSec AH header.

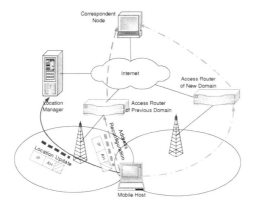

Figure 9.4: Interoperability between SIGMA and IPSec.

9.4 Summary

In this chapter, we described the issues of MIP interoperating with Ingress filtering. It is shown that SIGMA can interoperate with existing network security infrastructures such as Ingress filtering and can use IPSec to increase its robustness against connection hijack risks.

Chapter 10

Conclusions

We have presented SIGMA, $\underline{S}$eamless $\underline{IP}$-diversity-based $\underline{G}$eneralized $\underline{M}$obility $\underline{A}$rchitecture, to manage handovers of mobile nodes. SIGMA is an end-to-end mobility architecture, which means it does not require any change in Internet infrastructure. SIGMA can also inter-cooperate with existing Internet security mechanisms.

Through discrete time simulation, we compared the handover performance of SIGMA with three different Mobile IPv6 enhancements including FMIPv6, HMIPv6, and FHMIPv6. Different performance measures, including handover latency, packet loss and throughput, have been compared. Our results indicate that for typical network configuration and parameters, SIGMA has a *lower handover latency, lower packet loss rate* and *higher throughput* than the three MIPv6 enhancements. SIGMA has also been shown to be network *friendly* due to probing of the new network at every handover.

Using an analytical model, we have evaluated the signaling cost of SIGMA and compared with that of HMIPv6. Numerical results show that, in most practical scenarios, the signaling cost of SIGMA is lower than HMIPv6. Another analytical model is developed to evaluate the survivability of location management scheme used in SIGMA and compared with that of MIP. Numerical results have shown SIGMA has better survivability than Mobile IP in terms of system response time and service blocking probability, in practical environments under the risk of hardware failures and distributed DoS attacks.

150

To further reduce the signaling cost of SIGMA, a hierarchical location management scheme is also developed for SIGMA— HiSIGMA. This hierarchical scheme is also applicable to other transport layer mobility architectures. By introducing the concept of Anchor Zone Server into location management of mobile hosts, the signaling cost of SIGMA can be greatly reduced and is lower than that of HMIPv6.

We have also investigated the use of SIGMA in space networks to support IP mobility — SIGMA-SN. After analyzing different satellite handover scenarios in network-layer, we have shown that various components of SIGMA-SN can be directly mapped to the architectural elements of the space handover scenario. We have designed SIGMA in such a way that it will be applicable both in space networks and in Wireless LANs and cellular networks. This is in-line with NASA's objective of developing technologies which can be used in a wide range of applications to enable them to enjoy the research and development efforts of commercial vendors.

In summary, after considering various aspects of SIGMA, we believe SIGMA is well suited for IP mobility management in terrestrial and space mobile networks. We have validated the performance of SIGMA using simulation environment and real-world testbed. Ultimately, SIGMA will be used for managing both flat handovers between homogenous wireless networks and vertical handovers between heterogeneous networks.

Bibliography

[1] C.E. Perkins (editor), "IP Mobility Support," IETF RFC 3344, August 2002.

[2] D. Johnson, C.E. Perkins, and J. Arkko, "Mobility support in IPv6," IETF RFC 3775, June 2004.

[3] R. Koodli (editor), "Fast handovers for Mobile IPv6," IETF DRAFT, draft-ietf-mipshop-fast-mipv6-03.txt, October 2004.

[4] H. Soliman, C. Catelluccia, and K.E. Malki et al., "Hierarchical Mobile IPv6 mobility management (HMIPv6)," IETF DRAFT, draft-ietf-mipshop-hmipv6-04.txt, December 2004.

[5] R. Hsieh and A. Seneviratne, "A comparison of mechanisms for improving Mobile IP handoff latency for end-to-end TCP," in *ACM MobiCom*, San Diego, USA, September 2003, pp. 29–41.

[6] D. A. Maltz and P. Bhagwat, "MSOCKS: An architecture for transport layer mobility," in *INFOCOM*, San Francisco, USA, March 1998, pp. 1037–1045.

[7] A. C. Snoeren and H. Balakrishnan, "An end-to-end approach to host mobility," in *ACM MobiCom*, Boston, MA, August 2000, pp. 155–166.

[8] T. S. Rappaport, *Wireless Communications Principles and Practice*, Prentice Hall, Upper Saddle River, NJ, 1996.

[9] G. Caire, G. Taricco, and E. Biglieri, "Bit-interleaved coded modulation," *IEEE Transactions on Information Theory*, vol. 44, no. 3, pp. 927–946, May 1998.

[10] Matthias Holzbock, "IP based user mobility in heterogeneous wireless satellite-terrestrial networks," *Wireless Personal Communications*, vol. 24, no. 2, pp. 219–232, Jan 2003.

[11] J. Glossner, D. Iancu, J. Lu, E. Hokenek, and M. Moudgill, "A software-defined communications baseband design," *IEEE Communications Magazine*, vol. 41, no. 1, pp. 120–128, Jan 2003.

[12] S. Fu and M. Atiquzzaman, "SCTP: State of the art in research, products, and technical challenges," *IEEE Communications Magazine*, vol. 42, no. 4, pp. 64–76, April 2004.

[13] S. Fu, M. Atiquzzaman, and W. Ivancic, "Effect of delay spike on SCTP, TCP Reno, and Eifel in a wireless mobile environment," in *11th International Conference on Computer Communications and Networks*, Miami, FL, October 2002, pp. 575–578.

[14] S. Fu, M. Atiquzzaman, and W. Ivancic, "SCTP over satellite networks," in *IEEE 18th Annual Workshop on Computer Communications*, Dana Point, California, October 2003, pp. 112–116.

[15] M. Atiquzzaman and W. Ivancic, "Evaluation of SCTP multistreaming over wireless/satellite links," in *12th International Conference on Computer Communications and Networks*, Dallas, Texas, October 2003, pp. 591–594.

[16] G. Ye, T. Saadawi, and M. Lee, "SCTP congestion control performance in wireless multi-hop networks," in *MILCOM2002*, Anaheim, California, October 2002, pp. 934–939.

[17] M. Allman, V. Paxson, and W. Stevens, "TCP Congestion Control," IETF RFC 2581, April 1999.

[18] W. R. Stevens, *TCP/IP Illustrated, Volume 1 (The Protocols)*, Addison Wesley, November 1994.

[19] C.E. Perkins, "Mobile Networking Through Mobile IP," *IEEE Internet Computing*, vol. 2, no. 1, pp. 58–69, January/February 1998.

[20] E. Gustafsson, A. Jonsson, and C.E. Perkins, "Mobile IP regional registration," IETF DRAFT, draft-ietf-mobileip-reg-tunnel-04.txt, March 2001.

[21] R. Ramjee, T.L. Porta, and S. Thuel et al., "IP micro-mobility support using HAWAII," IETF DRAFT, draft-ietf-mobileip-hawaii-00.txt, June 1999.

[22] A. T. Cambell, S. Kim, and J. Gomez et al., "Cellular IP," IETF DRAFT, draft-ietf-mobileip-cellularip-00.txt, December 1999.

[23] K.E. Malki (editor), "Low latency handoffs in Mobile IPv4," IETF DRAFT, draft-ietf-mobileip-lowlatency-handoffs-v4-07.txt, October 2003.

[24] C.E. Perkins and K.Y. Wang, "Optimized smooth handoffs in Mobile IP," in *IEEE International Symposium on Computers and Communications*, July 1999, pp. 340–346.

[25] M.C. Jung, J.S. Park, D.M. Kim, H.S. Park, and J.Y. Lee, "Optimized hand-off management method considering micro mobility in wireless access network," in *5th IEEE International Conference on High Speed Networks and Multimedia Communications*, July 2002, pp. 182–186.

[26] I.W. Wu, W.S. Chen, H.E. Liao, and F.F. Young, "A seamless handoff approach of Mobile IP protocol for mobile wireless data networks," *IEEE Transactions on Consumer Electronics*, vol. 48, no. 2, pp. 335–344, May 2002.

[27] W. Liao, C.A. Ke, and J.R. Lai, "Reliable multicast with host mobility," in *IEEE Global Telecommunications Conference (GLOBECOM)*, November 2000, pp. 1692–1696.

[28] S. Fu and M. Atiquzzaman, "Improving end-to-end throughput of Mobile IP using SCTP," in *Workshop on High Performance Switching and Routing*, Torino, Italy, June 2003, pp. 171–176.

[29] R. Stewart and Q. Xie et. al., "Stream control transmission protocol," IETF RFC 2960, October 2000.

[30] R. Stewart and C. Metz, "SCTP: New transport protocol for TCP/IP," *IEEE Internet Computing*, vol. 5, no. 6, pp. 64–69, November/December 2001.

[31] A.L. Caro, J.R. Iyengar, and P.D. Amer et. al, "SCTP: a proposed standard for robust Internet data transport," *IEEE Computer*, vol. 36, no. 11, pp. 56–63, November 2003.

[32] A. Jungmaier, "Performance evaluation of the stream control transmission protocol," in *Proceedings of the IEEE Conference 2000 on High Perfomance Switching and Routing*, Heidelberg, Germany, June 2000, pp. 141–148.

[33] A. Jungmaier, E.P. Rathgeb, and M. Tuxen, "On the use of SCTP in failover-scenarios," in *International Conference on Information Systems, Analysis and Synthesis*, Orlando, Florida, July 2002, pp. 363–368.

[34] A.L. Caro, P.D. Amer, P.T. Conrad, and G.J. Heinz, "Improving multimedia performance over lossy networks via SCTP," in *ATIRP 2001*, College Park, MD, March 2001.

[35] S. Thomson and T. Narten, "IPv6 stateless address autoconfiguration," IETF RFC 2462, December 1998.

[36] R. Stewart, M. Ramalho, and Q. Xie et. al., "Stream control transmission protocol (SCTP) dynamic address reconfiguration," IETF DRAFT, draft-ietf-tsvwg-addip-sctp-09.txt, June 2004.

[37] B. Awerbuch and D. Peleg, "Concurrent online tracking of mobile users," in *ACM SIGCOMM Symposium on Communications, Architectures and Protocols*, September 1991, pp. 221–233.

[38] *The Network Simulator - ns-2*, http://www.isi.edu/nsnam/ns/.

[39] Robert Hsieh, Zhe Guang Zhou, and Aruna Seneviratne, "S-MIP: A seamless handoff architecture for Mobile IP," in *IEEE INFOCOM*, San Francisco, CA, April 2003, pp. 1774–1784.

[40] H. Chen and Lj. Trajkovic, "Route optimization in Mobile IP," in *Workshop on Wireless Local Networks (WLN)*, Tampa, FL, November 2002, pp. 847–848.

[41] C. Casetti and M. Meo, "A new approach to model the stationary behavior of TCP connections," in *IEEE INFOCOM 2000*, Tel-Aviv, Israel, March 2000, pp. 367–375.

[42] T.V. Lakshman and U. Madhow, "The performance of TCP/IP for networks with high bandwidth-delay products and random loss," *IEEE/ACM Transactions on Networking*, vol. 5, no. 3, pp. 336–350, June 1997.

[43] M. Mathis, J. Semke, and J. Mahdavi, "The macroscopic behavior of the TCP congestion avoidance algorithm," *Computer Communications Review*, vol. 27, no. 3, pp. 67–82, July 1997.

[44] J. Padhye, V. Firoiu, D.F. Towsley, and J.F. Kurose, "Modeling TCP Reno performance: a simple model and its empirical validation," *IEEE/ACM Transactions on Networking*, vol. 8, no. 2, pp. 133–145, April 2000.

[45] C. Casetti and M. Meo, "Modeling the stationary behavior of TCP Reno connections," in *International Workshop on Quality of Service in Multiservice IP Networks*, Rome, Italy, January 2001, pp. 141 – 156.

[46] Takayuki Osogami Adam Wierman and Jorgen Olsn, "A unified framework for modeling TCP-Vegas, TCP-SACK, and TCP-Reno," in *11th IEEE/ACM International Symposium on Modeling, Analysis and Simulation of Computer Telecommunications Systems*, Orlando, Florida, Octobor 2003, pp. 269–278.

[47] G. Bolch, S. Greiner, H. D. Meer, and K. S. Trivedi, *Queueing Networks and Markov Chains: Modeling and Performance Evaluation with Computer Science Applications*, John Wiley & Sons, New York, August 1998.

[48] M.Garetto, R. Cigno, M. Meo, and M.A. Marsan, "Closed queueing network models of interacting long-lived TCP flows," *IEEE/ACM Transactions on Networking*, vol. 12, no. 2, pp. 300–311, April 2004.

[49] K. S. Trivedi, *Probability and Statistics with Reliability, Queuing and Computer Science Applications*, John Wiley & Sons, New York, 2nd edition, October 2001.

[50] J. Xie and I. F. Akyildiz, "An optimal location management scheme for minimizing signaling cost in Mobile IP," in *IEEE International Conference on Communications (ICC)*, New York, April 2002, pp. 3313–3317.

[51] Y.W. Chung, D.K. Sun, and A.H. Aghvami, "Steady state analysis of P-MIP mobility management," *IEEE Communication Letters*, vol. 7, no. 6, pp. 278–280, June 2003.

[52] M.E. Crovella and A. Bestavros, "Self-similarity in world wide web traffic: evidence and possible causes," *IEEE/ACM Transactions on Networking*, vol. 5, no. 6, pp. 835–846, Dec 1997.

[53] V. Paxson and S. Floyd, "Wide area traffic: the failure of Poisson modeling," *IEEE/ACM Transactions on Networking*, vol. 3, no. 3, pp. 226–244, June 1995.

[54] C. Bettstetter, H. Hartenstein, and X. Perez-Costa, "Stochastic properties of the random waypoint mobility model: epoch length, direction distribution, and cell change rate," in *5th ACM International Workshop on Modeling Analysis and Simulation of Wireless and Mobile Systems*, September 2002, pp. 7–14.

[55] J. Xie and I. F. Akyildiz, "A novel distributed dynamic location management scheme for minimizing signaling costs in Mobile IP," *IEEE Transactions on Mobile Computing*, vol. 1, no. 3, pp. 163–175, July 2002.

[56] J.W. Lin and J. Arul, "An efficient fault-tolerant approach for Mobile IP in wireless systems," *IEEE Transactions on Mobile Computing*, vol. 2, no. 3, pp. 207–220, July-Sept 2003.

[57] T. You, S. Pack, and Y. Choi, "Robust hierarchical mobile IPv6 (RH-MIPv6): an enhancement for survivability and fault-tolerance in mobile IP systems," in *IEEE 58th Vehicular Technology Conference*, October Fall 2003, pp. 2014–2018.

[58] R. Jan, T. Raleigh, D. Yang, and L.F. Chang et. al., "Enhancing survivability of mobile Internet access using mobile IP with location registers," in *IEEE INFOCOM*, March 1999, pp. 3 – 11.

[59] R.Bush, D. Karrenberg, M. Kosters, and R. Plzak, "Root name server operational requirements," IETF RFC 2870, June 2000.

156

[60] J. Meyer, "On evaluating the performability of degradable computing systems," *IEEE Transactions on Computers*, vol. 29, no. 8, pp. 720–731, August 1980.

[61] J. Meyer, "Closed-form solutions of performability," *IEEE Transactions on Computers*, vol. 31, no. 7, pp. 648–657, July 1982.

[62] R.K.C. Chang, "Defending against flooding-based distributed denial-of-service attacks: a tutorial," *IEEE Communications Magazine*, vol. 40, no. 10, pp. 42 – 51, Oct. 2002.

[63] W. Xing, H. Karl, and A. Wolisz, "M-SCTP: Design and prototypical implementation of an end-to-end mobility concept," in *5th Intl. Workshop on the Internet Challenge: Technology and Applications*, Berlin, Germany, October 2002.

[64] S. J. Koh, M. J. Lee, M. L. Ma, and M. Tuexen, *Mobile SCTP for Transport Layer Mobility*, draft-sjkoh-sctp-mobility-03.txt, February 2004.

[65] S. Fu, L. Ma, M. Atiquzzaman, and Y. Lee, "Architecture and performance of SIGMA: A seamless mobility architecture for data networks," in *40th IEEE Internationl Conference on Communications (ICC)*, Seoul, Korea, May 2005.

[66] T. Pratt, C.W. Bostian, and J. Allnutt, *Satellite communications*, Wiley, New York, 2 edition, 2003.

[67] L. Wood, A. Clerget, and I. Andrikopoulos et. al., "IP routing issues in satellite constellation networks," *International Journal of Satellite Communications*, vol. 19, no. 1, pp. 69–92, January/February 2001.

[68] K. Bhasin and J. L. Hayden, "Space Internet architectures and technologies for NASA enterprises," *International Journal of Satellite Communications*, vol. 20, no. 5, pp. 311–332, September/October 2002.

[69] R. Donadio, F. Zeppenfeldt, and S. Pirio et. al. eds., *ESA Telecommunications: IP Networking over Satellite Workshop*, European Space Agency, Noordwijk, Netherlands, May 2004.

[70] K. Bhasin and J. L. Hayden, "Space Internet architectures and technologies for NASA enterprises," *International Journal of Satellite Communications*, vol. 20, no. 5, pp. 311–332, Sep 2002.

[71] J. Rash, E. Criscuolo, K. Hogie, and R. Praise, "MDP: Reliable file transfer for space missions," in *NASA Earth Science Technology Conference*, Pasadena, CA, June 11-13, 2002.

[72] NASA, "Omni: Operating missions as nodes on the internet," ip-inspace.gsfc.nasa.gov.

[73] K. Hogie, E. Criscuolo, and R. Parise, "Link and routing issues for Internet protocols in space," in *IEEE Aerospace Conference*, 2001, pp. 2/963–2/976.

[74] G. Minden, J. Evans, S. Baliga, S. Rallapalli, and L. Searl, "Routing in space based Internets," in *Earth Science Technology Conference*, Pasadena, CA, June 11-13, 2002.

[75] F. Hallahan, "Lessons learned from implementing Mobile IP," in *The Second Space Interent Workshop*, Greenbelt, MD, May 21-22, 2002.

[76] K. Hogie, "Demonstration of Internet technologies for space communication," in *The Second Space Internet Workshop*, Greenbelt, Maryland, May 21-22 2002.

[77] J. Rash, R. Casasanta, and K. Hogie, "Internet data delivery for future space missions," in *NASA Earth Science Technology Conference*, Pasadena, CA, June 11-13, 2002.

[78] K. Leung, D. Shell, W. Ivancic, D. Stewart, T. Bell, and B. Kachmar, "Application of Mobile-IP to space and aeronautical networks," *IEEE Aerospace and Electronic Systems Magazine*, vol. 16, no. 12, pp. 13–18, Dec 2001.

[79] M. Allman, D. Glover, and L. Sanchez, "Enhancing TCP over satellite channels using standard mechanisms," IETF RFC 2488, January 1999.

[80] C. Partridge and T.J. Shepard, "TCP/IP performance over satellite links," *IEEE Network*, vol. 11, no. 5, pp. 44 –49, Sep/Oct 1997.

[81] V. Jacobson, R. Braden, and D. Borman, "TCP extensions for high performance," IETF RFC 1323, May 1992.

[82] N. Ghani and S. Dixit, "TCP/IP Enhancements for Satellite Networks," *IEEE Communications Magazine*, vol. 37, no. 7, pp. 64–72, July 1999.

[83] A.C. Clarke, "Extra-terrestrial relays," *Wireless World*, vol. LI, no. 10, pp. 305–308, Oct 1945.

[84] J.L. Grubb, "Iridium overview - the traveler's dream come true," *IEEE Communications Magazine*, vol. 29, no. 11, pp. 48–51, Nov 1991.

[85] C.E. Fossa, R.A. Raines, G.H. Gunsch, and M.A. Temple, "An overview of the IRIDIUM low earth orbit (LEO) satellite system," in *IEEE National Aerospace and Electronics Conference (NAECON)*, July 1998, pp. 152 – 159.

[86] H.N. Nguyen, S. Lepaja, J. Schuringa, and H.R. Vanas, "Handover management in low earth orbit satellite IP networks," in *GlobeCom*, Nov. 2001, pp. 2730–2734.

[87] B. Sarikaya and M. Tasaki, "Supporting node mobility using mobile IPv6 in a LEO-satellite network," *International Journal of Satellite Communications*, vol. 19, no. 5, pp. 481–498, September/October 2001.

[88] Y.H. Kwon and D.K. Sung, "Analysis of handover characteristics in shadowed LEO satellite communication networks," *International Journal of Satellite Communications*, vol. 19, no. 6, pp. 581–600, November/December 2001.

[89] S. Dixit, "Wireless IP and its challenges for the heterogeneous environment," *Wireless Personal Communications*, vol. 22, no. 2, pp. 261–273, August 2002.

Appendix A

Acronyms

3G 3rd Generation Wireless System

AH Authentication Header

AR Access Router

AP Access Point

AZS Anchor Zone Server

CMR Call to Mobility Ratio

CN Correspondent Node

CCoA Collocated Care of Address

CoA Care of Address

CTMC Continuous Time Markov Chains

DDoS Distributed Denial of Service

DHCP Dynamic Host Configuration Protocol

DNS Domain Name Server

ESP Encapsulating Security Payload

FA Foreign Agent

FMIPv6 Fast Handovers for Mobile IPv6

FHMIPv6 Fast and Hierarchical Mobile IPv6

FQDN Fully Qualified Domain Name

HA Home Agent

HMIPv6 Hierarchical Mobile IPv6

HOL Head-Of-Line

HZS Home Zone Server

IETF Internet Engineering Task Force

IP Internet Protocol

IPSec IP Security

LM Location Manager

MAP Mobility Anchor Point

MH Mobile Host

MIP Mobile IP

MIPv6 Mobile IPv6

MRT Mobile Routing Table

MTTF Mean Time To Failure

MTTR Mean Time To Repair

PCS Personal Communications Services

PKI Public Key Infrastructure

RMMP Reliable Mobile Multicast Protocol

SAA Stateless Address Auto-configuration

SACK Selective Acknowledgment

SCTP Stream Control Transmission Protocol

SIGMA Seamless IP-diversity-based Generalized Mobility Architecture

SMR Session to Mobility Ratio

TCP Transmission Control Protocol

VoIP Voice over IP

WLAN Wireless Local Area Network

Appendix B

Implementation of SIGMA in *ns*-2

In order to investigate the advantage of SCTP multihoming based handover scheme in *ns*-2 simulator, two interfaces have to be used for a wireless node. The problems arising from the current *ns*-2 simulator are multi-fold:

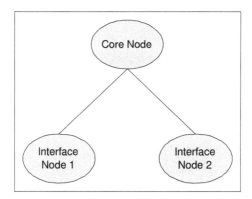

Figure B.1: Existing SCTP multihomed node structure in *ns*-2.

(1) The SCTP implementation for *ns*-2 by University of Delaware using a multihoming structure as shown in Fig. B.1, in which a multihomed node is a composite one composed of a core node and two interface nodes. The core node and interface nodes are normal *ns*-2 nodes connected by a wired link. This kind of configuration cannot be used in mobile environments since the

composite node cannot move around. We need to modify the wireless node structure to make the node multihomed by itself.

(2) The SCTP layer code supporting the multihoming in the SCTP implementation by the University of Delaware depends on the node structure of Fig. B.1. So, we also need to modify the SCTP layer code to accommodate the new multihomed wireless node structure.

(3) The addressing scheme of *ns*-2 only supports one IP address for each node. When the structure of the mobile nodes is modified, we hope each mobile node can have more than one IP address to be used by the transport layer protocols.

(4) The routing protocols and ARP protocol in *ns*-2 only support single-homed nodes. When the structure of the mobile nodes is modified, these protocols can't route packets correctly to the nodes with specified IP address or resolve the IP address to the MAC address. These protocols need to be modified accordingly to support the multihoming infrastructure.

This chapter is organized as follows. First, the overview of the implementation is described in Appendix B.1. Then, the modifications I have made in the node structure, network layer and transport layer will be described in Appendix B.2, B.4, and B.5 respectively. In Appendix B.6.2, we show a simulation script which have been used to test the modifications I have made at the different layers.

B.1 Implementation Overview

We have modified the *ns*-2 wireless node structure as shown in Fig. B.2. The normal part is the old structure of a wireless node, and the shaded ones are the added interface corresponding to physical, MAC, and LLC layer functions which are generally implemented by a wireless interface card. For modifying the SCTP patch so that we can handle multiple interfaces and multiple IP addresses in a

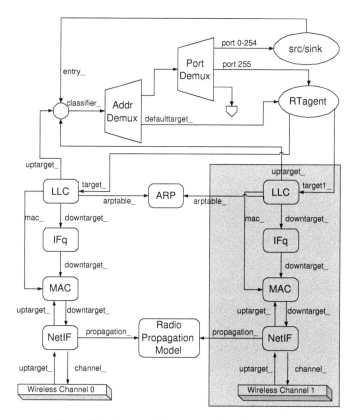

Figure B.2: New multihomed mobile node structure.

wireless mobile environment, the *ns*-2 simulator is modified at five different layers: physical, MAC, LLC, network, and transport layer. Note that the physical, MAC, and LLC layers are part of the node structure as shown in Fig. B.2. The objects grouped by the dashed box are inserted part into the existing node structure.

To utilize the new *ns*-2 mobile node structure, we also need to change routing agent implementation at the network layer to make the simulator support routing to the transport layer agent attached to this kind of multihomed nodes. Two kinds of routing protocols are used, NOAH for base stations and MhNOAH for mobile nodes. The NOAH protocol is developed by UC Berkeley to support routing in mobile IP environment, while the MhNOAH protocol is a newly developed protocol in this project to deliver the packets to and to send the packets from the mobile nodes through current active interface.

A set of new classes are developed to exchange location management information between the mobile nodes, home agent, and foreign agent. This is required when a third party user want to connect to a mobile node when it has moved out of the coverage of the home agent, since the user must know from the home agent where the mobile node are. The new classes are implemented as Agent/MhBS and Agent/MhMH to be used at base stations and mobile nodes respectively.

As for the ARP protocol, the old ARP module provided by ns-2 only support one MAC address in one node, this will make the base stations connected to the multihomed mobile nodes can't find the second MAC address on the mobile nodes, and the packets will be dropped at the link layer. The ARP module is re-engineered to get IP address information from link layer and populate it into the ARP table to support multihomed mobile node address resolution.

Finally, at the transport layer, the SCTP implementation for ns-2 by University of Delaware uses a multihoming structure that can't support mobile node operations. A new class is implemented in ns-2 simulator: Agent/SCTP/MhSCTP based on Delaware group's SCTP patch. This class will cooperate with the multihomed mobile node structure as well as the multihomed routing protocol, and use current routing table to set the source and destination address for outgoing packets. This approach will provide a more natural support for SCTP multihoming feature than the SCTP implementation provided by University of Delaware.

166

B.2 Modifications to Mobile Node Structure

The *ns*-2 wireless node structure was modified, and a new class was implemented in *ns*-2 simulator: Node/MobileNode/MhMobileNode (Mh means multi-homed).

B.2.1 Related Source Files

(1) New files: `$NS/mhmobilenode.cc`, `$NS/mhmobilenode.h`, and `$NS/tcl/lib/ns-mhmobilenode.tcl`

(2) Modified files: `$NS/tcl/lib/ns-lib.tcl`, `$NS/tcl/lib/ns-mip.tcl`, and `$NS/tcl/lib/ns-default.tcl`

B.2.2 Objective of Modifications

(1) `mhmobilenode.cc` and `mhmobilenode.h` are used to implement Node/MobileNode/MhMobileNode object in C++ scope.

(2) `ns-mhmobilenode.tcl` is used to plumb existing *ns*-2 objects into a multi-homed mobile node and attach a MhSCTP (will be introduced in Sec. B.5) agent on the node.

(3) `$NS/tcl/lib/ns-lib.tcl`, `$NS/tcl/lib/ns-mip.tcl`, and `$NS/tcl/lib/ns-default.tcl` are modified to make the new node type can co-operate with existing *ns*-2 modules.

B.2.3 Program Flow

The program flow chart for creating multihomed mobile nodes is shown in Fig. B.3. Shadowed boxes are simulation user interface, the other boxes are called TCL instprocs or C++ functions. The normal line style means a flow in the simulation script while the dashed lines mean the relation between modules. The file names are the locations for the instprocs or functions.

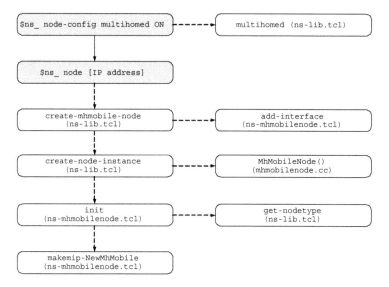

Figure B.3: Program flow for creating multihomed mobile nodes.

B.2.4 Functions in `mhmobilenode.cc`

- `MhMobileNode(void)`: constructor function of the class.

- `virtual int command(int argc, const char*const* argv)`: used as an interface function to the simulation script, implement some commands that can be used during the simulation.

- `void add_interface(void)`: used to add interface/IP address dynamically during simulation, currently is implemented as a null function.

- `void delete_interface(void)`: used to delete interface/IP address dynamically during simulation, currently is implemented as a null function.

B.2.5 Instance Procedures in `ns-mhmobilenode.tcl`

- `init`: constructor function, accept zero, one or more IP addresses as arguments to enable the `MhMobileNode` object use multiple IP addresses. It will also call instproc `makemip-NewMhMobile` to config a registering agent into the node to support mobile IP operations.

- `reset`: reset `NETIF`, `MAC`, `LL`, `IFQ`, `ARPTABLE` objects within the mobile node.

- `makemip-NewMhMobile`: create a registering agent to register mobile node with the home agent and foreign agent, and config this agent into the node structure to support mobile IP operations.

- `add-interface`: plumb a new interface into the node structure given the following arguments: channel (channel) , radio propagation model (`pmodel`), link layer type (`lltype`), MAC type(`mactype`), Queue type (`qtype`), Queue length (`qlen`), Interface type (`iftype`), and Antenna type (`anttype`). The argument channel should be provided by simulation script, other arguments can get their default values from file `ns-default.tcl` or been configured by simulation script.

- `mhattach`: attach a `MhSCTP` (will be introduced in Sec. B.5) agent on the multihomed mobile node. Set source addresses (`agent_addr_`, `agent_addr1_`) that agent will use to fill the IP header in the packets sent by this agent; allocate a new port and assign it to the `MhSCTP` agent.

B.2.6 Instance Procedures in `ns-lib.tcl`

- `multihomed`: put variable `multihomed_` in Simulator's space to identify whether the new created nodes should be multihomed mobile nodes.

- `get-nodetype`: set node type to `MhMobile` if the node is multihomed mobile node.

- node-config: keep consistent with *ns*-2 node configuration style. Set variable multihomed_ in Simulator's space if user issue a command like: $ns node-config multihomed ON.

- node: main interface with simulation script to create new nodes. If multihomed_ is ON, call instproc create-mhmobile-node to create new multihomed mobile node.

- create-mhmobile-node: call instproc create-node-instance to create a general node instance, then add two interfaces by calling instproc add-interface in ns-mhmobilenode.tcl to make the node composed of multiple interfaces.

- create-node-instance: If multihomed_ is ON, call constructor instproc init in ns-mhmobilenode.tcl to create new multihomed mobile node, and get multiple IP addresses.

B.2.7 Instance Procedures in ns-mip.tcl

- For all instprocs that need to judge whether the current node class is Node/MobileNode, also add a statement to judge whether the current node class is Node/MobileNode/MhMobileNode.

B.2.8 New Line in ns-default.tcl

- Simulator set multihomed_ OFF: By default set the node type as single-homed.

B.3 Modification to Introduce Layer 2 connection Setup and IP Address Resolution Latencies

In the state of the art mobile system technologies, when a mobile host changes its point of attachment to the network from one wireless network to another, it needs to perform a Layer 2 (data link layer) handover to cutoff the association with the old access point and re-associate with a new one. As an example, in IEEE802.11 WLAN infrastructure mode, this Layer 2 handover will require several steps: detection, probe, and authentication and reassociation with new AP. Only after these procedures have been finished, higher layer protocols can proceed with their signaling procedure, such as Layer 3 router advertisements. In the case of SIGMA, since each MH is equipped with two interface cards, a Layer 2 connection setup instead of handover is carried out.

Once the MH finishes Layer 2 connection setup and receives the router advertisement from the new access router (AR2), it should begin to obtain a new IP address . This task can be accomplished through several methods: DHCP, DHCPv6, or IPv6 stateless address auto-configuration. We call the time required for MH to acquire the new IP address as address resolution time.

The MAC layer we used for implementation is IEEE 802.11 WLAN protocol. In the current *ns*-2 implementation, no Layer 2 connection setup latency and address resolution latency is implemented. No signaling messages are exchanged for Layer 2 setup and new IP assignment. The Layer 2 handover (Layer 2 connection setup) is assumed to be performed seamlessly since MH can received the packets from both access routers as long as it stays in the overlapping region. Besides, the IP addresses are pre-assigned before the simulation (in case of Mobile IP) or obtained immediately from first router advertisements (in case of SIGMA). The objective of the code modification is to introduce Layer 2 handover/setup latency and address resolution latency into *ns*-2 so that we can show the advantage of SIGMA over Mobile IP and its enhancements in terms of handover latency and packet loss rate.

B.3.1 Introduction of Layer 2 Beacons

The standard IEEE 802.11 MAC layer in *ns*-2 does not provide a beacon mechanism to enable us detecting the entry and exit of the radio range of an access point (AP). We adopted a module which implements the PCF mode of the IEEE 802.11

wireless LANs in *ns*-2. It allows a node to become Point Coordinator (PC), and let that node send beacons, initiate Contention Free Periods (CFPs), and poll other stations during these CFPs in order to give higher priorities to such stations. The authors of the module are A. Lindgren and A. Almquist with LULEÅ University Of Technology, Sweden. We only need the beacon mechanism for our job, and do not need to use the full functions provided in this module such as the polling and CFPs.

B.3.2   Related Source Files

(1) `$NS/mac.h`

(2) `$NS/mac-802_11.h`

(3) `$NS/mac-802_11.cc`

(4) `$NS/mac-timers.h`

(5) `$NS/mac-timers.cc`

B.3.3   Implementation of Layer 2 Handover Latency

- Add a link list structure for caching existing APs using structure (mac/mac-802_11.h): _APList

- Define two additional timers (`mac/mac-timers.h`)
  - class `L2HandoffTimer`; (for L2 handover latency timing)
  - class `ExpCheckerTimer`; (for checking AP expiration when MH moves out of range)

- Timer handler for the two new timers (`mac-802_11.cc`)
  - `virtual void L2HandoffHandler(void)`
  - `virtual void ExpCheckerHandler(void)`

- Discard data packets when `L2HandoffTimer` is pending (`mac-802_11.cc`)

- Void Mac802_11::sendDATA(Packet *p) for outgoing data
- Void Mac802_11::recv_timer() for incoming data

- Define one additional command L2-latency (mac-802_11.cc)

  int Mac802_11::command(int argc, const char*const* argv)

B.3.4 Program Flow

The program flow chart for layer 2 connection setup/handover latency implementation is shown in Fig. B.4. The shadowed boxes are simulation user interface, the other boxes are TCL instprocs or C++ functions. The file names are the locations for the instprocs or functions.

B.3.5 Implementation of Address Resolution Latency

The address resolution latency does not require a separate implementation since it has same effect on upper layer protocols as Layer 2 setup/handover latency. Therefore, if the address resolution latency is needed, just add the value of this latency into Layer 2 setup/handover latency and simulate it using the interface we discussed in Sec. B.3.3.

B.4 Modifications to Network Layer

Current *ns*-2 uses node address (basically the node id) to route packets to an agent. If we use multiple interfaces (multiple IP addresses), the way used to identify the agent has to be changed because one SCTP agent may attached to multiple IP addresses, and only using one node address is not sufficient. Also, the mobile nodes need to exchange information with base stations to perform location management. I have created three new classes: Agent/MhNOAH_Agent, Agent/MhBS and Agent/MhMH in *ns*-2 simulator to make it compatible with multiple IP addresses of a mobile node.

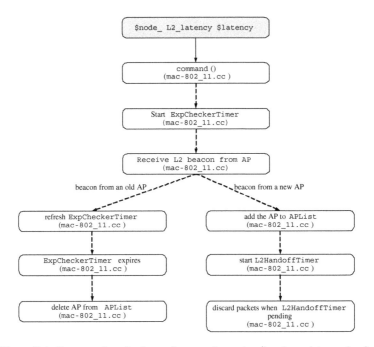

Figure B.4: Program flow for Layer 2 connection setup/handover latency implementation.

B.4.1 Related Source Files

(1) $NS/mhmobilenode.cc

(2) $NS/mhreg.cc

B.4.2 Objective of Modifications

Make the routing agent can handle multiple IP addresses, and perform the location management for the mobile nodes.

B.4.3 Functions in `mhmobilenode.cc`

- `void forwardPacket (Packet *p)`: forward packets based on current active interface addresses instead of node address. In this function, we need to find which interface the mobile node are using to exchange information to the base stations, then using this information to set the next hop address, and forward it.

- `void sendoutBCastPkt (Packet *p)`: send advertisement by base stations and channel requests by mobile nodes.

- `void recv (Packet *p, Handler *)`: main entry point of the class, dispatch packets to different functions based on packet type.

- `int command (int argc, const char* const* argv)`: interface to the user simulation scripts.

B.4.4 Functions in `mhreg.cc`

- `MhBSAgent()`: constructor for MhBS class.

- `MhMHAgent()`: constructor for MhMH class.

- `int command (int argc, const char * const * argv)`: interface to the simulator TCL scripts.

- `void send_ads(int dst, NsObject *target)`: send advertisement from base stations.

- `virtual void recv(Packet *, Handler *) *)` : main entry point of the class, dispatch packets to different functions based on packet type.

- `int reg(AgentList*)`: perform the registration by the mobile nodes.

- `void send_sols()`: send channel requests by mobile nodes.

B.5 Modifications to Transport Layer

The SCTP implementation for *ns*-2 by University of Delaware using a multihoming structure that can't support mobile node operations. In Secs. B.2 and B.4, we described the modifications to the *ns*-2 wireless node structure and routing agent implementation to make the simulator support multihomed node and routing to the transport layer agent attached to this kind of nodes. In this section, the modifications made to the existing *ns*-2 SCTP module, and the new implemented class Agent/SCTP/MhSCTP is introduced.

B.5.1 Related Source Files

(1) New files: `$NS/mhsctp.cc`, `$NS/mhsctp.h`, and `$NS/tcl/lib/ns-mhsctp.tcl`

(2) Modified files: `$NS/sctp.cc` and `$NS/tcl/lib/ns-default.tcl`

B.5.2 Objective of Modifications

:

(1) `mhsctp.cc` and `mhsctp.h` are used to implement `Agent/SctpAgent/MhSctpAgent` object in C++ scope.

(2) `ns-mhsctp.tcl` is used to plumb existing *ns*-2 objects into a multihomed mobile node and attach a `MhSCTP` agent on the node.

(3) `ns-default.tcl` are modified to set default values for additional agent source and destination addresses and ports (`agent_addr1_`, `agent_port1_`, `dst_addr1_` and `dst_port1_`).

B.5.3 Program Flow

The program flow chart for `MhSCTP` agent implementation is shown in Fig. B.5. The shadowed boxes are simulation user interface, the other boxes are TCL instprocs or C++ functions. The file names are the locations for the instprocs or functions.

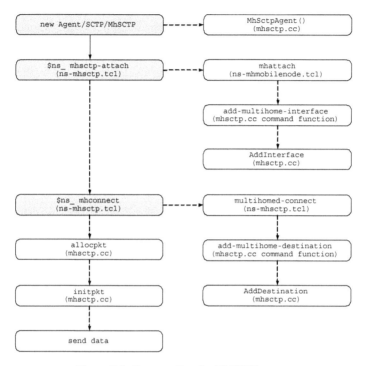

Figure B.5: Program flow for MhSCTP agent.

B.5.4 Functions in `mhsctp.cc`

- `MhSctpAgent(void)`: constructor function of the class.

- `~MhSctpAgent(void)`: destructor function of the class.

- `virtual void delay_bind_init_all(void)`: Init the binding of TCL variables `agent_addr1_`, `agent_port1_`, `dst_addr1_` and `dst_port1_` to C++ variables. The binding of other TCL variables for agent address and port (`agent_addr_`, `agent_port_`, `dst_addr_` and `dst_port_`) has been done by agent.cc.

- virtal void delay_bind_dispatch(void): Cooperate with delay_bind_init
  _all(void) and perform the actual binding of TCL variables agent_addr1_,
  agent_port1_, dst_addr1_, dst_port1_ to C++ variables.

- virtual int command(int argc, const char*const* argv): interface function with simulation script, implement some commands that can be used
  during the simulation, including: reset, close, set-primary-destination,
  force-source, add-multihome-destination and add-multihome-interface.

- void AddInterface(int iNsAddr, int iNsPort): MhSctpAgent maintain
  a linked list of all interfaces, each node of the link list consist of the interface address, port assigned, and the pointer to the next node and previous
  node. This function will create a new node in the linked list, and assign
  iNsAddr, iNsPort to the related fields in the node. The function will be
  called by add-multihome-interface in the mhsctp-attach instproc to notify a MhSCTP agent about new available interfaces and their addresses.

- virtual void AddDestination(int iNsAddr, int iNsPort): MhSctpAgent
  maintain a linked list of all destinations, each node of the link list consist of
  the destination address, port assigned, and the pointer to the next node and
  previous node. This function will create a new node in the linked list, and assign iNsAddr, iNsPort to the related fields in the node. The function will be
  called by add-multihome-destination in the simulation to notify a MhSCTP
  agent about new available destination addresses.

- virtual void SetPrimary(int iNsAddr): set current primary destination
  as iNsAddr. The function will be called by set-primary-destination in the
  simulation to instruct a MhSCTP agent using specified destination address to
  send data or perform retransmissions.

- virtual void ForceSource(int iNsAddr): set current primary interface
  as iNsAddr. The function will be called by force-source in the simulation

to instruct a MhSCTP agent using specified interface to send data or perform retransmissions.

- void initpkt(Packet* p, ns_addr_t src_, ns_addr_t dest_) const: Fill in the packet (pointed by Packet* p) with specified source address (addr_) and destination address (dest_), this function will override the function initpkt (Packet* p) implemented by agent.cc, which can only uses the default source and destination addresses.

- Packet* allocpkt(ns_addr_t src_, ns_addr_t dest_) const: Allocate memory for a new packet and call function initpkt to fill in the packet with specified source address (addr_) and destination address (dest_). This function will override the function Packet* allocpkt (void) implemented by agent.cc.

- Packet* allocpkt(ns_addr_t src_, ns_addr_t dest_, int n) const: Call allocpkt(ns_addr_t src_, ns_addr_t dest_) to allocate a new packet then fill the packet payload with n bytes of data.

  - ns_addr_t SetSource(void) set the source address of the outgoing packets from this agent.
  - ns_addr_t SetDestination(SctpDest_S *)set the primary destination address to be handled by the routing protocol.

B.5.5 Instance Procedures in ns-mhsctp.tcl

- mhsctp-attach: call instproc mhattach of MhMobileNode class to attach an MhSCTP agent on a MhMobileNode node. Maintain a TCL list named multihome_bindings_ to record all the addresses and ports available for local interfaces.

- mhconnect: connect source agent and destination agent to setup a SCTP association. At least one side of the agents should be MhSCTP agent.

- `multihomed-connect`: called by `mhconnect`, for each side of source and destination agent, this function calls `add-multihome-destination` to notify the other side about local interface information.

- `mhsimplex-connect`: called by `mhconnect`, set `dst_addr_` and `dst_addr1_` of each side agent of the association as the `agent_addr_` and `agent_addr1_` of the other side.

B.5.6 New Line in `ns-default.tcl`

Initialize TCL variables `agent_addr1_`, `agent_port1_`, `dst_addr1_`, `dst_port1_` to C++ variables. The initialization of other TCL variables for agent address and port (`agent_addr_`, `agent_port_`, `dst_addr_` and `dst_port_`) has been done by original `Agent` class.

(1) `Agent/SCTP/MhSCTP set agent_addr1_ -1`

(2) `Agent/SCTP/MhSCTP set agent_port1_ -1`

(3) `Agent/SCTP/MhSCTP set dst_addr1_ -1`

(4) `Agent/SCTP/MhSCTP set dst_port1_ -1`

B.6   SIGMA Simulation by Example

We have described the modification we have made to the mobile node structure, network,and SCTP layers. In this section we describe the TCL script which we have been using to test the modifications and simulate the performance of SIGMA.

B.6.1   Simulation Topology

The TCL script simulates the network shown in Fig. B.6. The mobile host (MH) can communicate with AR1 and AR2 by two separate wireless channels. MH will begin to move from AR1 to AR2 at time 5.0 sec, it will move out of the coverage of AR1 at time 16.0 sec.

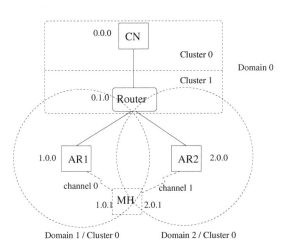

Figure B.6: Example simulation topology.

B.6.2 Sample User Simulation Script

```
#create two wired nodes, W(0) for correspondent node and W(1) for router


set W(0) [$ns_ node 0.0.0]
set W(1) [$ns_ node 0.1.0]


#configure wireless network properties
set opt(chan)          Channel/WirelessChannel  ;# channel type
set opt(prop)          Propagation/FreeSpace     ;# radio-propagation model
set opt(netif)         Phy/WirelessPhy           ;# network interface type
set opt(mac)           Mac/802_11                ;# MAC type set
opt(ifq)               Queue/DropTail/PriQueue   ;# interface queue type
set opt(ll)            LL                        ;# link layer type
set opt(ant)           Antenna/OmniAntenna       ;# antenna model
set opt(ifqlen)        50                        ;# max packet in ifq
set opt(nn)            1                         ;# number of mobilenodes
set opt(adhocRouting)  NOAH                      ;# routing protocol
set chan_1_            [new $opt(chan)]          ;# create channel 1
```

```
set chan_2_            [new $opt(chan)]           :# create channel2


# Configure for wireless node properties
$ns_ node-config       -multihomed OFF            ;# turn off the multihomedproperty
                       -mobileIP OFF              ;# turn off MIP functions
                       -adhocRouting $opt(adhocRouting)
                       -llType $opt(ll)
                       -macType $opt(mac)
                       -ifqType $opt(ifq)
                       -ifqLen $opt(ifqlen)
                       -antType $opt(ant)
                       -propType $opt(prop)
                       -phyType $opt(netif)
                       -channel  $chan_1_
                       -topoInstance $topo
                       -wiredRouting ON
                       -multihomed OFF
                       -agentTrace ON
                       -routerTrace ON
                       -macTrace ON


# Create AR1 and AR2
#use chan_1_ for AR1
set AR1 [$ns_ node 1.0.0]
#use chan_2_ for AR2
$ns_ node-config -channel $chan_2_
set AR2 [$ns_ node 2.0.0]
# turn AR1 and AR2 into a Point Coordinator
[$AR1 set mac_(0)] make-pc
[$AR2 set mac_(0)] make-pc
# set layer 2 beacon period
[$AR1 set mac_(0)] beaconperiod 10ms
[$AR2 set mac_(0)] beaconperiod 10ms


# Configure for MH node
```

```
$ns_ node-config        -multihomed ON              ;# turn on the multihomedproperty
                        -wiredRouting OFF           ;# no wired routing on mobile host
                        -adhocRouting MHNOAH        ;# use MHNOAH routing for mobile nodes
                        -channel $chan_1_           ;# channel 1 as primary channel
                        -channel1 $chan_2_          ;# channel 2 as secondary channel

# create a mobile node that would be moving between AR1 and AR2
set MH [$ns_ node 1.0.1 2.0.1]                      ;# MH have two IP addresses
# set the both interfaces of MH have an layer 2 setup latency of 200ms
[$MH set mac_(0)] L2-latency   200ms
[$MH set mac_(1)] L2-latency   200ms

# create a MhSCTP agent and attach it to MH
set src [new Agent/SCTP/MhSCTP]
$ns_ mhsctp-attach $MH $src
$src set debugFileIndex_ 0
$src set dataChunkSize_   512
$src set mtu_ 576
$src set numOutStreams_ 1
$src set initialSsthresh_   65536
$src set initialRwnd_   [expr [$src set dataChunkSize_] * 20 ]
$src set initialSsthresh_   [expr [$src set dataChunkSize_] * 20 ]
$src set initialCwndMultiplier_ 2
$src set useMaxBurst_ 1
$src set fid_ 1
$ns_ color 1 "blue"

# create a normal SCTP agent and attach it to CN
set sink [new Agent/SCTP]
$sink set debugFileIndex_ 1
$sink set dataChunkSize_   512
$sink set mtu_ 576
$sink set initialRwnd_   [expr [$sink set dataChunkSize_] * 20 ]
$sink set initialSsthresh_   [expr [$sink set dataChunkSize_] * 20]
$sink set useDelayedSacks_ 0
```

```
$ns_ attach-agent $W(0) $sink
# connect src and sink agents
$ns_ mhconnect $src $sink

#create the FTP application and start data transmission
set ftp [new Application/FTP]
$ftp attach-agent $src
$ns_ at 0.0"$ftp1 start"

#mobile host begin moving from AR1 to AR2 at time 5 seconds
$ns_ at 5.000000000000 "$MH setdest 840.0 300.0 30.0"
$ns_ run
```

Wissenschaftlicher Buchverlag bietet

kostenfreie

Publikation

von

wissenschaftlichen Arbeiten

Diplomarbeiten, Magisterarbeiten, Master und Bachelor Theses
sowie Dissertationen, Habilitationen und wissenschaftliche Monographien

Sie verfügen über eine wissenschaftliche Abschlußarbeit zu aktuellen oder zeitlosen
Fragestellungen, die hohen inhaltlichen und formalen Ansprüchen genügt,
und haben **Interesse an einer honorarvergüteten Publikation**?

Dann senden Sie bitte erste Informationen über Ihre Arbeit per Email
an info@vdm-verlag.de. Unser Außenlektorat meldet sich umgehend bei Ihnen.

VDM Verlag Dr. Müller Aktiengesellschaft & Co. KG
Dudweiler Landstraße 125a
D - 66123 Saarbrücken

www.vdm-verlag.de

www.ingramcontent.com/pod-product-compliance
Lightning Source LLC
LaVergne TN
LVHW022314060326
832902LV00020B/3448